# TECHNOLOGY AND UNDERDEVELOPMENT

# Technology and Underdevelopment

FRANCES STEWART

*First published 1977 by*
THE MACMILLAN PRESS LTD
*London and Basingstoke*
*Associated companies in New York*
*Dublin Melbourne Johannesburg and Madras*

ISBN 0 333 19510 8

Typeset, printed and bound
in Great Britain by
REDWOOD BURN LIMITED
Trowbridge & Esher

*This book is sold subject*
*to the standard conditions*
*of the Net Book Agreement*

To Michael

# Contents

# Preface

While working on the topics discussed in this book I have received help and support from various sources. The Rockefeller Foundation provided some financial assistance for the surveys (Chapter 9 and 10) in Kenya in 1969. The Ministry of Overseas Development supported the research for three years, 1969–72. I am grateful to the O.D.M. for this and for their approval of publication in this form. I have received extremely useful research assistance at different times from Gareth Jenkins, Jill Rubery and Jeffrey James. I am particularly grateful to Jeffrey for his assiduous help over the past few months. Numerous people have commented on various papers, which, in revised form, have become chapters in this book. I should particularly like to mention Charles Cooper (for comments on Chapter 5), G. K. Helleiner and participants at the Uppsala Conference in September 1974 (Chapter 7), and Ajit Bhalla (Chapter 10). For encouragement and stimulus I am above all indebted to Paul Streeten. Finally, as always, Michael Stewart's sceptical wit provided both challenge and support.

Three of the chapters are versions (with many minor and some major amendments) of papers that have been published elsewhere: Chapter 6 appeared in *Employment, Income Distribution and Development Strategy: Essays in Honour of H. W. Singer*, ed. A. Cairncross and M. Puri (Macmillan, 1976); Chapter 7 is based on an essay in G. K. Helleiner (ed.), *A World Divided* (Cambridge University Press, 1976); and Chapter 10 appeared in A. Bhalla (ed.), *Technology and Employment in Industry: A Case Study Approach* (Geneva: I.L.O., 1975).

Some of the discussion in the first part of this book has its origin in two earlier papers: 'The Choice of Technique in Developing Countries', *Journal of Development Studies*, October 1972, and 'Technology and Employment in LDCs', *World Development*, March 1974.

F.S.

*May 1976*

# Introduction

The majority of people in third world countries are very poor. Yet these countries possess – in abundant supply – the most important factor of production – manpower. The widespread poverty goes hand in hand with misuse of this critical resource – much labour is idle for much of the time, while when in use labour productivity is very low. The technology used, broadly identified with the methods of production in use, must then be at the heart of any analysis of questions of poverty and development in the third world. The third world gets almost all its technology from the advanced countries:[1] it is technologically dependent on the advanced countries where most of the search for new technology takes place.

The main aim of this book is to explore the impact of technology on development – and in particular to consider the connections between poverty and maldistribution of income in poor countries, and their condition of technological dependence.

The technology available determines the boundaries of what it is possible for a country to do. An economy may, of course, fail fully to exploit all the possibilities open to it: it can operate inside these outer boundaries, but it cannot step outside them. Over time the technology available alters, and so consequently do the boundaries it sets.

For the most part, economists have tended to make two assumptions about technology: first, that it is to be regarded as a parameter of the system, as far as economics is concerned – determined by scientists and technologists independently of economic facts. Consequently, the task of economists is to make the best of a bad job – i.e. to see how the best decisions may be made, within the boundaries set by technological possibilities. Secondly, that technical change is neutral – with respect to class and country and factor use – so that the boundaries represented by technological knowledge, which change over time in response to new scientific and technical developments, do so in a neutral way: any distortions in development patterns resulting from the use of technology must then be attributed to *misuse* rather than the nature of technology itself.[2]

This view of technology is inconsistent with the facts of economic development in poor countries. It is apparent there that the technology adopted is distorting the pattern of development, favouring a particular lop-sided pattern of growth, creating employment problems and skewed income distribution, and failing to make use of poor countries' natural and human resources.

The facts of economic development in poor countries are thus in conflict with the theoretical treatment of technology in much of economics. In considering the impact of technology on development one is also unavoidably drawn into examination of the theoretical treatment. This examination, and attempts to formulate alternative approaches, form a second theme to the book. The two themes – the substantive (i.e. consideration of how technology is affecting development patterns), and the theoretical (i.e. consideration of the adequacy of the treatment of these matters in economic theory) – are intertwined and discussed simultaneously throughout the book.

Chapter 1 considers the nature of technological choice: the historical determinants of the total technology available at any time, and the determinants of the choice made from within the total available, here described as the selection mechanisms. The normal assumptions of neutral and exogenous technical change are challenged both by the facts of technological development, as highlighted by poor countries' experience, and by consideration of the historical conditions of technological development. Chapter 1 presents a critique of conventional theory, and suggests a more general model of technological choice.

The question of the use – or rather mal-use and lack of use – of manpower is fundamental to analysis of development problems. Chapter 2 reviews recent approaches to the definition, nature and origin of employment problems in poor countries, and considers how the analysis relates to questions of technology. Chapter 2 also presents a dynamic view of employment problems, showing how in many countries trends over time soon dwarf any analysis of employment problems based on static measures.

Chapter 3 describes the characteristics of technology emanating from the advanced countries, in the light of the earlier discussion of the determinants of technological change. It looks at how the characteristics of that technology are influenced by the economic and social conditions in advanced countries, and shows how these characteristics have distorting effects on development in poor countries, where economic and social conditions differ. The result, it is argued, is that the technology third world countries get from rich countries is inappropriate.

Chapter 4 discusses the nature of *appropriate* technology – its defining characteristics, and the obstacles which inhibit the pursuit of an appropriate technology strategy. These obstacles include lack of scientific and technical research, the difficulties – technically and economically – of introducing an appropriate technology into a system based on the use of inappropriate technology and obstacles created by the political economy of technological choice.

Chapter 5 takes the concept of technological dependence as its central theme. It discusses the nature, extent and consequences of technological dependence, the way in which technological dependence relates to the more general treatment of the dependent status of poor countries, and considers alternative strategies for reducing technological dependence. In one sense, as already emphasised, the whole of the book is concerned with the question of technological dependence of poor countries on rich countries. Chapter 5 tries to draw together and consider arguments particularly related to dependence, but it does not itself present a complete picture of technological dependence, which must be considered in the light of the book as a whole.

Chapter 6 looks at the role of capital goods industries in poor countries, first in relation to the theories advanced to justify the build-up of heavy industry in Russia and India, and secondly in the light of the technology argument.

Chapter 7 considers the role of international trade, with particular emphasis on the direction of trade as between rich and poor nations (North/South trade), and within groupings of nations of similar income (North/North and South/South). The chapter includes discussion of different theories of trade, and institutional changes that might be conducive to an improved orientation of trade patterns.

These chapters (2–7) are chiefly concerned with macro-aspects of technology and underdevelopment. But the macro-picture is formed by many micro-decisions. The next three chapters of the book focus on the micro-decisions.

Chapter 8 is divided into two parts: the first half looks at the sort of problem encountered in micro-studies of choice of technique. The second half reviews the results of recent studies, attempting to fit the studies into the general framework of technological choice presented in Chapter 1, and to shed some light on the critical question of the existence and efficiency of alternative appropriate technology.

Chapter 9 describes a case study of choice of technique in maize grinding in Kenya.

Chapter 10 reports on a case study of cement-block manufacture in Kenya.

Finally, Chapter 11 briefly draws together some conclusions.

NOTES

1. The various terms to describe *rich* and *poor* countries, such as *advanced* and *less developed*, *developed* and *underdeveloped*, *North* and *South*, as normally used, have particular connotations, related to the economic/political views of the user. The terms are used more or less interchangeably in this book. It is hoped that they will be regarded purely as *proper names*: description and views are explicit in the text and not to be derived from nomenclature.

2. This does give a highly oversimplified view of economists' treatment of technical change. See, for example, Kaldor (1957) on the effects of the rate of investment on technical change, Arrow (1962) on learning by doing, Schmookler (1966) on economic influences on patented inventions, Salter (1969) on technical change, Atkinson and Stiglitz (1969) on localised technical change. However, despite these, and others, most analysis continues to treat technical change as exogenous and neutral.

# 1 The Technological Choice

This book is about the impact of technology on development. That impact depends on the technology in use in underdeveloped countries, which is a function of the technology available to the country, and the choice made from the total available. This chapter is concerned to discuss the nature of the technological choice.

Technology is often identified with the hardware of production – knowledge about machines and processes. Here a much broader definition is adopted, extending to all the 'skills, knowledge and procedures for making, using and doing useful things'.[1] Technology thus includes methods used in non-marketed activities as well as marketed ones. It includes the nature and specification of what is produced – the product design – as well as how it is produced. It encompasses managerial and marketing techniques as well as techniques directly involved in production. Technology extends to services – administration, education, banking and the law, for example – as well as to manufacturing and agriculture. A complete description of the technology in use in a country would include the organisation of productive units in terms of scale and ownership. Although much of the discussion will be in terms of technological development in the hardware of technology, the wider definition is of importance since there are relationships between the hardware and the software – between, for example, mechanical process and managerial techniques and infrastructural services – which help determine the choice made in both spheres.

Technology consists of a series of techniques. The technology available to a particular country is all those techniques it knows about (or may with not too much difficulty obtain knowledge about) and could acquire, while the technology in use is that subset of techniques it *has* acquired. It must be noted that the technology available to a country cannot be identified with all known techniques: on the one hand weak communication may mean that a particular country only knows about part of the total methods known to the world as a whole. This can be an important limitation on technological choice. On the other hand, methods may be known but they may not be available because no one is producing the machinery or other inputs required. This too limits technological choice.

Each technique is associated with a set of characteristics. These characteristics include the nature of the product, the resource use – of machinery, skilled and unskilled manpower, management, materials and energy inputs – the scale of production, the complementary products and services involved etc. Any or all of these characteristics may be important in determining whether it is possible and/or desirable to adopt a particular technique in a particular country and the implications of so doing.

More formally, we may think of all the known techniques as $wT = \{Ta, Tb, Tc, Td, \ldots Tn\}$ (where 'known' means known to the world) as constituting world technology. For a particular country, the technology available for adoption is that subset of world technology known to the country in question *and* available. Say, $cT = \{\bar{T}a \ldots \bar{T}n\}$, where $c$ denotes the country and the bar indicates that only techniques known to the country and available are included. Thus $cT \subset wT$.

Each of the techniques *Ta*, *Tb* . . . etc. is a vector consisting of a set of characteristics, ai, aii, aiii, bi, bii, biii . . . Thus technology can be described in matrix form, with each column representing the characteristics of each technique, as follows:

Matrix of World Technology = $wT$

| *Characteristics* | *Ta* | *Tb* | *Tc* | *Td* | *Te* |
|---|---|---|---|---|---|
| Product type | | | | | |
| Product nature | | | | | |
| Scale of production | | | | | |
| Material inputs | | | | | |
| Labour input: | | | | | |
| skilled | | | | | |
| unskilled | | | | | |
| Managerial input | | | | | |
| Investment requirements | | | | | |

The technology in use in a particular country is that subset of the technology available to it that has been selected and introduced, or $uT = \{\bar{T}a \ldots \bar{T}n\}$ where $uT \subset cT \subset wT$.

The processes by which world technology is narrowed down to an actual set of techniques in use may be crudely described as follows:

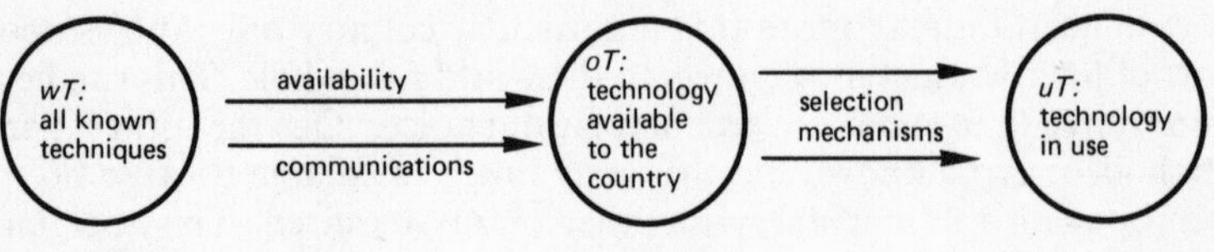


FIG. 1.1

The actual technology in use is thus circumscribed first by the nature of world technology, then by the availability to the country of known techniques, and finally by the choice made among those available. If the technology in use is thought to be inappropriate, it may be inappropriate because world technology is inappropriate, or because an inappropriate subset is available to the country, or because an inappropriate selection is made, or for some combination of the three reasons. Confusion is caused by failing to distinguish between the three. The rest of this chapter analyses technological choice for underdeveloped countries in terms of these three stages. Finally the more conventional approach to choice of technique is contrasted with the approach adopted here.

## WORLD TECHNOLOGY

Techniques do not exist in heaven, in Platonic caves or in entrepreneurs' imagination, ready to be plucked from the air and incorporated into use. Techniques, whether they be methods of administration or machines to produce consumer goods, have to be invented, developed, introduced, modified, etc. The development of techniques is essentially a historical process in which one technique with one set of characteristics replaces another in the light of the historical and economic circumstances of the time. The historical nature of technological development means that the time and circumstances in which any particular technique is developed heavily influence its characteristics. In the first place, scientific and technical knowledge varies between places and, especially, over time, Such variation over time has been particularly marked in the now industrialised countries over the last two hundred or so years.[2] The state of scientific and technical knowledge thus sets the scene, or provides the starting point, for new developments. Because of the fairly steady scientific and technical developments during this time, it is generally true that the later techniques have some advantage compared with earlier in terms of this starting point, and for this reason are likely to have greater productivity and efficiency in relation to resource use.

While the state of science may set the scene, other historic/economic circumstances are of critical importance in determining the characteristics of the techniques introduced. Techniques are only developed and introduced if they are believed to be viable in the economy in which they are introduced. This viability requires that the resource use involved fits in with the resource availability – in terms of nature of resources used, quantity and price; it also requires that there is an adequate market for the products produced. In a profit-maximising perfectly competitive economy one can be more precise about the necessary conditions for successful innovation – viz., given the ruling prices of resources and products, the new technique will generate more than normal profits. In the more real world of market imperfections, and varied motivation among innovators, it is still true that a new innovation in a capitalist economy must fit sufficiently well with resource costs and markets to produce long-run profits. Whether or not a technique is introduced and developed is thus firmly anchored in the historic and economic conditions of the date at which it is introduced. For example, the twentieth century saw the development of techniques for the mass production of cars. Such developments drew heavily on technical developments that were not available at a previous time; on the resource side, the developments required the mass use of materials, such as rubber and steel, of a type and quality that had only recently been developed; it also required massive investment that twentieth-century levels of income permitted. Education of the labour force that occurred in the late nineteenth century produced a work force suited to these production methods. On the demand side, mass markets were forthcoming as a result of the increase in incomes.

Economic and historic circumstances are themselves the product of the technology in use in a society, and hence of past technical developments. This should be clear from the car example: the availability of particular materials at sufficiently low prices was due to past innovations in rubber and steel production; the high incomes which provided the savings to finance the massive investment were the result of the use of high-productivity technology; similarly, the markets for cars also arose from the higher wages and salaries produced by the (relatively) high-productivity technology already in use. Thus technical innovation and economic change are interrelated, with one feeding off the other, often making it difficult to distinguish which is cause and which effect.

It is perhaps helpful to classify these economic/historic circumstances which condition the characteristics of techniques into three categories, and discuss each one briefly:

(1) organisation of production
(2) income levels (and distribution)
(3) technical factors

(1) *Organisation of production:* clearly methods designed for use in family enterprises are likely to differ from those intended for production on a mass scale by a multinational enterprise, with say 1000 people employed in each plant, specialisation and flexible movement of materials, parts, and management between plants, often in different countries, and sometimes in different continents. Where family enterprise predominates technology is likely to be small-scale and personal. Specialisation is likely to be less with a large proportion of any product produced within the family. In contrast, the multinational enterprise may specialise on a large scale in each plant, and assemble the parts in a different plant. The remoteness of boss from workers – indeed the difficulty of identifying a boss – may lead to depersonalisation of the productive process. These are extreme examples, but illustrate the general point that the

nature of the organisation of production affects the design of technology. To take a different example, a system in which a single state enterprise has the monopoly over energy production is likely to lead to a different type of energy technology than one in which each enterprise produces its own energy. There is a two-way relationship between technology and economic organisation: technical developments may lead to changes in methods of organisation, or may accentuate trends. For example, the change to a factory system in the eighteenth and nineteenth century was partly due to the economies of concentrating sources of power – partly but not entirely, since technical developments might have occurred that enabled continued putting-out and self-employment; that they did not, has been attributed to socio-economic changes that were occurring independently of the technical change.[3] The existing organisation of production conditions technical developments, particularly when the technical developments are themselves initiated and carried out within or for these units. This is true of much research and development today which is the product of the research departments of large companies.[4]

Scale of productive units, specialisation, method of organisation of work, and managerial techniques are obvious examples of characteristics of techniques which are much influenced by how the economy is organised.

(2) *Income levels*: income levels are of significance from the resource side and the market side. On the resource side, the overall availability of savings is largely determined by average income levels. Savings ratios do vary according to the distribution of incomes and other factors, but the key determinant remains the income level. The availability of savings in turn determines the average investment per member of the workforce. Hence, as incomes rise, the ratio of investment to employee also tends to rise. This also applies to expenditure on educating the workforce and improving its health. Techniques introduced into higher-income societies may normally take for granted more educated and healthier labour forces. The opportunity cost of labour also changes with incomes: higher-income societies are higher-income because of the greater productivity of the labour force. This greater productivity means that the opportunity cost of labour (and the wage-rate) are higher. In economies with generally high labour productivity, only techniques involving high labour productivity will be introduced because only they will be profitable. Low-productivity techniques (relative to the average in the economy) will not be able to finance sufficient wages to pay for the necessary labour.

On the market side, the income level (and its distribution) determines the size and nature of the market. Engel pointed out that as incomes rose the proportions of expenditure on different categories of good changed; in particular, expenditure on food fell as a proportion. Probably of greater significance, the nature of products consumed within each broad category changes.

As people (and societies) get richer they do not simply consume the same as before but more of it – ten bags of potatoes a year instead of two: the nature of the products they consume also changes. The same broad needs are fulfilled by a different set of products, embodying a different (on the whole more satisfying, more sophisticated) set of characteristics, with higher standards. Rice tends to be replaced by wheat and by meat; medieval 'players' by movies and by radio; radio is replaced by television, first black-and-white and then colour. Cars take over from vehicles powered by horse or humans, and the cars become more sophisticated, faster and quieter. Changes in the nature of products consumed are an essential aspect of getting richer: long before modern science and technology made possible the sort of product replacement described above, richer societies and people were distinguished from their poorer contemporaries by the nature of the goods they consumed, as well as by their quantity – by their silks and their palaces. Modern science and technology made possible various products which were previously unknown, even unthinkable, and it put the search for new products on a systematic basis. These new products are both cause and consequence of higher incomes: higher incomes provide the purchasing power and hence the markets which make it worth developing new products; and the technological developments make possible the mass production of sophisticated goods, embodying new materials, which are the basis of the increases in income.

The nature of the product, which is an important characteristic of techniques, is thus related to the income levels of the consumer for whom they are designed. Product specifications are also related to cultural factors and to the requirements imposed by the technology in use: for example, if the technology used to weave requires a particular type and standard of thread this may largely determine the characteristic of the thread.

(3) *Technical factors*: this is shorthand to describe the technical requirements imposed by the system as a whole. Any single technical innovation has to fit in with the rest of the system both in terms of the requirements it imposes for inputs, and in terms of the demand for the good. The thread case just mentioned is an example of requirements imposed by one part of the technology in use on other parts. A new technique must use inputs that are available, or can be made available, and must provide output which will fit into further production if it is an intermediate good, or into consumption patterns if it is a consumer good. While these requirements leave some leeway for variations, they also impose restrictions. There are technological linkages between different parts of the system which mean that much of technology comes as a package, which cannot be separated and introduced bit-by-bit, but which goes together.[5] The requirements of a technique extend beyond the material inputs directly involved in the productive process to

managerial inputs and infrastructural services. Thus the efficient use of a particular technique may only be consistent with sophisticated managerial methods involving advanced methods of accounting, and computerised stock control; it may impose particular demands for energy, water and transport.[6] Technical requirements extend to methods of administration in the system as a whole; the type of law and order required for successful operation, the tax system, etc., are all related to the technology in use. Levels of living[7] of the labour force may be another technical requirement. The required labour input, in terms of energy, concentration, punctuality and literacy are related to the technology. All these requirements depend on the levels of living – and therefore the real incomes – of the workers and on their education. This is not to argue that each technique imposes a unique set of requirements, and can only be operated if these requirements are met. Clearly, there is some variation possible in terms of the quality and quantity of most of the inputs mentioned. But any variation tends to lead to variations in the productivity of the techniques, and sufficient deviation from the sort of inputs for which the techniques are designed may lead to a total breakdown. For example, cars designed for advanced-country roads, assuming the availability of highly trained mechanics for maintenance and repair, will work on murram roads, as the East African safari shows. But they will not work where there are no roads at all, and the length of life and efficiency of operation will be seriously affected by the different conditions in countries with roads which are not tarmacked and with few mechanics.

Techniques are developed against a background of a particular technology package – a particular set and quality of inputs, infrastructural services, legal/administrative system, labour force, etc. Some of the techniques may impose new requirements on the system, leading to the development of new or improved inputs, or, for example, retraining of the labour force. But even where this is so, they tend to take for granted more than they change. A technological innovation generally forms a ripple in the pool of technology in use, not a hurricane, or a tornado. Thus the date and place of the introduction of the technique determines the nature of the technology package it is designed to go with.

In the industrial societies where most technical development takes place, technological and economic developments form an interconnected and self-reinforcing dynamic cycle. As incomes rise wages rise, making existing technology uneconomic compared with labour-saving innovations. Rising incomes allow greater expenditure on investment per employee. The later techniques combine greater scientific and technical knowledge with greater investment expenditure and consequently tend to be associated with rising labour productivity, leading to a further rise in incomes and a further incentive for innovation. There is also a dynamic cycle in product development: innovations largely consist in new and/or improved products;[8] these new products in part constitute (and are required by) the rise in per capita incomes and consumption levels. Technical requirements also form a dynamic cycle: each innovation changes the technology package in use, and hence the technical requirements further innovations may assume and require. The innovation may impose requirements that are not fully met by the system, and hence provide opportunity and incentive for further innovations.[9] These innovations in turn may assume a certain type of technology package, change it and impose new demands for its successful use. For example, changed accounting techniques may involve computerisation: once introduced on any scale this provides an incentive for improved computer techniques, which in turn may lead to other changes in managerial methods, including computerised payments and stock control. Simultaneously, the innovations impose requirements for changes in training requirements; once achieved, the new training permits other innovators to take for granted a particular type of trained labour force, and design new techniques against this background.

Aspects of the cycles of technological development are illustrated below:

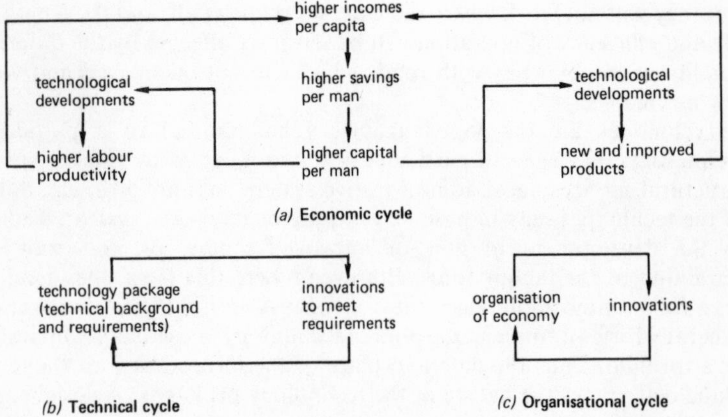


FIG. 1.2 Technological developments in a (dynamic) closed system
*Arrows indicate the direction of causality*

In the top diagram, (*a*), three interconnecting cycles have been distinguished. First, on the left, that of increases in per capita incomes providing an incentive for technological developments which lead to higher productivity, thus making possible further increases in incomes. Secondly, on the right, higher incomes provide an incentive for technological developments involving new and improved products, while these new products in turn are an essential aspect of further increases in incomes. Thirdly, in the centre, higher incomes lead (with constant propensity to save) to higher savings per man and consequently greater capital accumulation per man, thus influencing and making possible technological developments of both types, techniques and products. Cycle (*b*) illustrates the technological development cycle in terms of technical requirements, with the existing technology in use both providing the background of techniques in use, which an innovation may take for granted, and imposing demands for changes which an innovation may fulfil. Innovations made against this background in turn change the technology package – the background against which further innovations are made. These innovations take the organisational framework more or less as a datum, but they may reinforce or conflict with existing organisational forms and thus contribute to change in organisational forms. Whether they reinforce or conflict with existing forms may be determined partly by technological factors, and partly by who makes the innovation: for example innovations introduced by the communes in China are likely to be suitable for operation at a commune level, while innovations made by a large-scale organisational unit are likely to be consistent with operation at large scale. Yet technical factors *may* dictate the form the innovations take: for example, small-scale iron and steel manufacture at commune level was found to be technically inefficient, thus ruling out iron and steel manufacture at this level (though much depends on how 'efficiency' is defined and how many resources are put into the development of techniques at different levels). All three cycles take place simultaneously and are different ways of describing the same process of innovation. They cannot be disentangled. It is virtually impossible to distinguish between product and process changes, since one man's product is another man's process; most changes in process simultaneously change the end product, while most product changes require process changes for their realisation. Each technical innovation has to fit into each of the cycles – the economic cycle of factor payments, income and savings levels, the technical cycle and the organisational one.

Technology available today is the consequence of the historical process in which technology evolves. The total of all techniques ever developed encompasses the whole historical process. The total may thus be viewed as consisting of a succession of techniques, developed at different times, say *T*1800, *T*1801, *T* . . . *T*1975, where *T* represents the techniques developed in the year described by the time subscripts. The characteristics of the technique depend as argued on the historic circumstances in the place where it was introduced – the techniques have to fit into the particular stage of the dynamic cycles that the economy has reached. Because most technological development, since the industrial revolution, has occurred in the developed countries, we may start by assuming that *T*1800 . . . refers to techniques developed in and for the developed countries. With this assumption, world technology today depends on the characteristics of the techniques developed in the advanced countries, which in turn depends on the historic/economic circumstances of the advanced economies during this period, which conditioned their dynamic cycles. Broadly speaking, the period has been characterised by systematic changes: these changes include rising incomes, increasing labour productivity, increased education of manpower, increased range and sophistication of products designed to fit in with rising incomes, increasing specialisation, increased investment per employee, a factory system of production which has increasingly, along with employer/employee work relationships, drawn in the whole society, and so on. Such developments will be discussed in more detail and with some empirical backing in Chapter 3.

In so far as countries may choose between techniques of different vintages, they also choose between a different set of characteristics associated with techniques of different vintages. However, many of the earlier techniques become obsolete over time – indeed since new techniques are often *designed* to replace previous techniques, rather than to complement them, this is scarcely surprising. The obsolescence factor reduces the choice far below the total *T*1800 . . . How far below depends on the source of obsolescence, and on the countries' ability to reproduce techniques now obsolete in advanced countries.

### OBSOLESCENCE

Obsolescence occurs with economic developments and technical change, and is the product of the interaction between the two. There are many sources of obsolescence: generally speaking, the relevance of the source of obsolescence varies according to the nature of the economy. Some sources of obsolescence tend to make the technique obsolete in any society, while others only apply in so far as other changes are occurring, and may be applicable to advanced countries and not to underdeveloped countries. In categorising sources of obsolescence, it must be remembered that while the nature of the source of obsolescence varies, particular technical changes may embody or reflect, in part or in whole, *all* the sources, so that selectivity about which sources to accept and which to reject is not then possible. It is none the less useful to start by trying to disentangle sources of obsolescence. For analytical purposes, therefore, we start by distinguishing between obsolescence of methods of production to produce a given product, and product obsolescence. For each, a distinction may be made between obsolescence associated with changes in the economic structure – obsolescence which would not occur without such changes; and obsolescence associated with increases in efficiency that make previous techniques obsolete irrespective of economic changes. In addition, there is obsolescence independent of economic changes and of changes in the efficiency of substitute techniques, such as occurs with changes in the technology in use in the rest of the system. (*a*) *Obsolescence of methods of production to produce given products*: changes in the economic structure affect the sort of technique it is profitable to develop and introduce, and may then cause older techniques to become obsolete. One example that tends to receive much emphasis is changes in relative prices. For example, suppose, as shown in the diagram below, there are two resources used, *m* and *n*, to produce a given product. Assume the initial method of production is shown as 1800 in the diagram, and the initial price of the resources *PmPn*. Now assume that the price of the resources changes to *P′mP′n*. The 1800 technique will remain profitable so long as it is the only technique, but the new prices combined with general developments in knowledge may lead to a new technique, 1810, being developed. At the old prices the 1810 technique would have been unprofitable, as compared with the old technique; at the new prices, the 1800 technique becomes obsolete. Thus technical changes combined with economic changes (in this case the relative price of resource *m* and resource *n*) cause obsolescence.

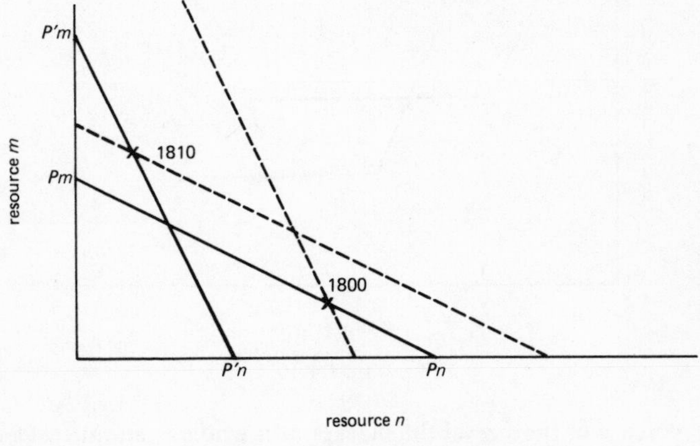


FIG. 1.3

This type of change is *reversible* in the sense that were the prices to change back to the original prices, it would be worth reverting to the original technique. Similarly, countries with different prices might find it worth while going back to the 1800 technique. The exposition has deliberately been in general terms, of resource *m* and resource *n*, although it is most often interpreted as a matter of investment resources and labour resources, with new techniques tending to use more investment resources in relation to labour, while increasing availability of investment resources and reduced price (relative to labour) provide the associated economic changes. Changes in the price and availability of different types of skilled labour, of various materials (e.g. wood,[10] coal, oil) have been associated with this type of obsolescence. Availability can be as important, many believe more important, than price. If resource *m* runs out altogether, or is only available in very limited quantities, it may cause any technique using much of it to become obsolete.

Another type of change in the economic structure associated with obsolescence is that of changes in the scale of production. Techniques may be competitive at one scale,[11] obsolete at another. Suppose there are two techniques *A* and *B*, depicted in Figure 1.4, with technique *B* designed for a larger scale than technique *A*. Assume that at and above the scale for which the technique is designed, costs per unit of output are constant, but below that scale they rise sharply with diminishing capacity utilisation. Then at a scale of output greater than 0*a* technique *B* is more efficient, and technique *A* becomes obsolete; at a scale less than 0*a* technique *A* is more efficient.

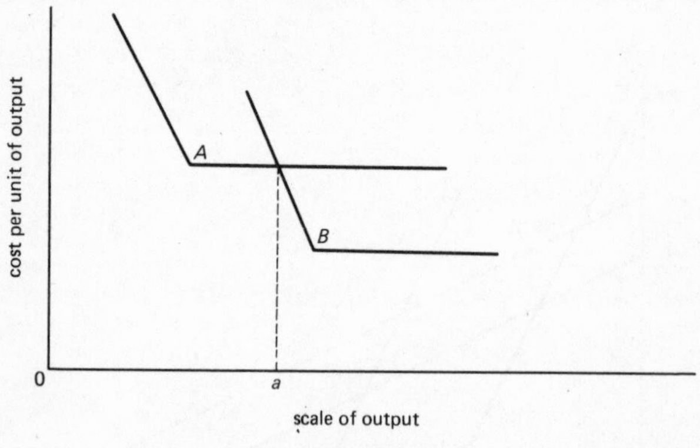


FIG. 1.4

An increase in the size of the market as a whole is an intrinsic aspect of economic growth[12]. Historically, in the advanced countries the typical size of plant has grown faster than the market as a whole.[13] Technological developments, consequently, have taken the form of increased scale – i.e. a movement from *A* to *B* on the diagram. Obsolescence associated with increased scale may be reversible – i.e. for smaller markets technique *A* again becomes the more profitable. But though reversible in this sense, they are not reversible in the same way as changes in relative factor prices, because, once developed, technique *B* offers lower unit costs than technique *A*, and therefore presents an incentive for expansion of the market so that the economies may be realised.

Technical change may produce new techniques that are more efficient than old, *irrespective* of the economic structure. Figure 1.5 shows a technique developed in 1820 which uses less of resource *m*, and no more of resource *n*, than the old technique. Consequently, irrespective of the relative price of resources *m* and *n*, the old technique becomes obsolete. The technique of 1800 is now defined as technically inefficient as compared with the technique of 1820.[14]

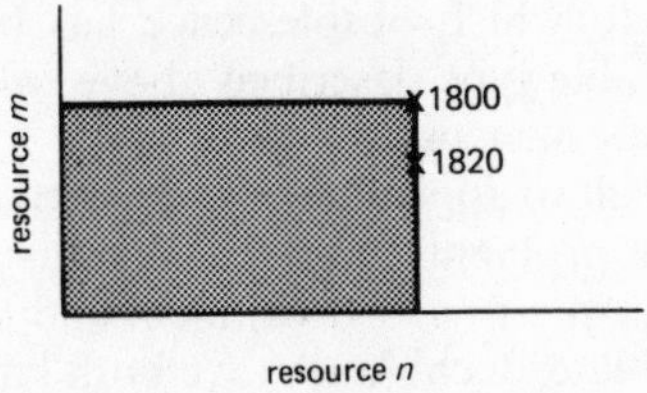


FIG. 1.5

The shaded area in the diagram shows the area in which new technical developments would make the 1800 technique technically inefficient.

Technical advances may occur in jumps so that over a period techniques are rendered obsolete through changed factor payments, as discussed above, and then some drastic advance is made, rendering all previous advances technically inefficient. Figure 1.6 illustrates, with *T*1800, *T*1810, *T*1820 to *T*1850 all representing changed factor use, the previous techniques becoming inefficient economically due to changed factor prices, but none the less remaining possibly efficient techniques

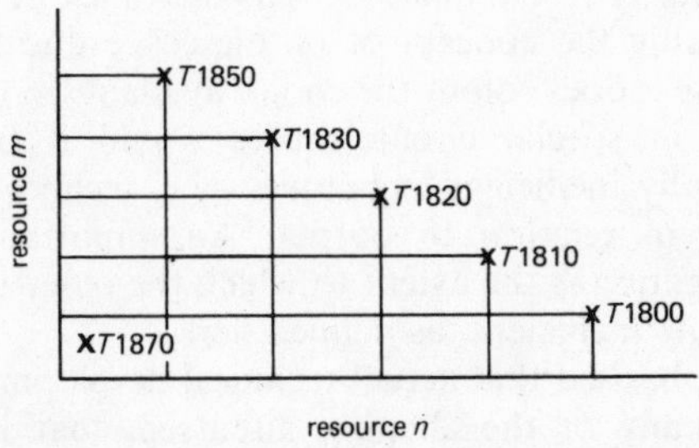


FIG. 1.6

for countries with different factor prices. Then a new advance is made, *T1870*, which involves such savings in resource use that all previous techniques become technically inefficient.

Obsolescence due to changes which make earlier techniques inferior in this way is not reversible, and applies to any economy, developed or underdeveloped, which regards both resources as scarce, and wishes to maximise output in relation to scarce resources. Thus the nature of the technological choice before underdeveloped countries is much influenced by the extent to which obsolescence has been of this type, as opposed to the reversible type described above, which is responsive to differences in economic structure.

However, it is difficult to apply this model to actual technical developments because of the problems involved in measuring resources. The obvious application is in terms of investment and labour requirements. It appears that many later techniques save both labour and investment in relation to output, and would therefore seem to be changes of the type described (with later techniques making earlier ones technically inefficient or inferior). However, whether (and the extent to which) this is true depends on how investment is measured – since investment resources are not homogeneous, their measure requires some system of weighting.[15] Similarly, though perhaps less critically, labour is not homogeneous, and while the total number of labourers declines, the skill content rises. Hence the labour-saving element also depends on the particular set of weights used to 'add' labour of different types. Once measurement becomes dependent on the weighting adopted, whether or not a particular technique is obsolete does depend on the economic structure and is not independent of it. It is thus very difficult to classify actual technical developments into those that are and those that are not dependent on the economic structure. Moreover, technically inefficient techniques may be worth selecting for their distributive consequences, if these distributive consequences are not obtainable in any other way. Thus techniques that are apparently technically inefficient may none the less not always be obsolete.

Despite these fairly severe qualifications that must be borne in mind, it is worth pursuing the concept of obsolescence due to technical inefficiency, because it does colour the choice available to underdeveloped countries. Only in special circumstances would it be worth while selecting technically inefficient techniques – i.e. techniques using more of *all* resources in relation to output. An important question for empirical investigation is the extent to which the older techniques have become technically inefficient, as defined here.

It must be emphasised that actual technical developments simultaneously embody many of the changes discussed, that is they tend to reflect the changes in the economic structure, being larger-scale, more investment-intensive, as well as being of greater productivity, as they were developed at a later stage of scientific and technical development. (*b*) *Product obsolescence*: this is one of the most important sources of obsolescence in techniques: the technique is rendered obsolete because demand for the product – of the particular design and quality – falls off. Products may be categorised into consumer goods and intermediate goods (including parts for machines, etc.), though of course some products may be both final and intermediate goods. Sources of product obsolescence differ between the two. As far as intermediate goods are concerned, changes in demand are due to changes in the technology in use in the rest of the system. Thus if a particular technique becomes obsolete, for one of the reasons discussed above, demand for the products – the machines used, material inputs etc. – that go to make it disappears and is replaced by demand for the products involved in the technique which has replaced it. Changes in demand for final consumption goods depend on how consumer tastes change – which is a function of income levels and distribution, of product promotion, and of technical advances. Changes in income levels have a critical effect on the nature of consumption demand, in respect of product design, quality and quantity, as argued above. Thus products become obsolete over time because of income changes with people switching to higher-quality goods as they become richer. 'Producers' sovereignty'[16] also influences the nature of consumption demand, with heavy promotion of new products making existing products obsolete.[17] Finally, scientific and technical advances may make products more efficient, just as they make processes more efficient. Products can be regarded as (normally) indivisible bundles of characteristics which fulfil various needs.[18] For example, roof thatching fulfils the need for protection from rain, insulation, and possibly also contributes to aesthetic needs and to social standing. A particular product may fulfil needs well or badly, completely or incompletely, and may be combined with other desired or undesired consequences: for example, a typewriter may function efficiently in terms of getting words on to paper, but it may produce ugly print, and be noisy – another typewriter might be quiet, but scarcely legible, or stiff to operate, and so on. If new products are described as more efficient than old, this means that they fulfil the needs for which they are demanded more successfully (completely/efficiently) and/or have fewer undesirable side-effects at no greater resource use. We have just argued that a more efficient fulfilment of needs, and the fulfilment of a greater variety of needs, is an important aspect of how products change as incomes rise, and a major reason why products become obsolete with rises in income. The replacement of the noisy hand-operated typewriter by the swift silent electric typewriter is an example. In theory, one can distinguish between these changes in products which involve additional need fulfilment at no greater resource cost (and/or some resource saving), and those which increase need fulfilment *and* resource cost. We may define the former as changes leading to greater product efficiency, so that previous products become obsolete because they become inefficient as compared with the new products, and the latter as higher income products. For example, a new transistor which is at least as cheap as the older version, and is also lighter to carry and gets a better reception, is described as a more efficient product, but a stereo system which also achieves greater need fulfilment but involves greater resource cost is a higher-income product. This theoretical distinction is difficult to apply in practice because of ambiguity about the meaning and measurement of *resource cost* and of *need fulfilment*. Because each product normally fulfils more than one need, a change in products may mean that some needs are fulfilled better, others worse, compared with the previous situation, and the net effects of the change are very much a matter of judgement. It is then impossible to say unambiguously that the new product has more/less need fulfilment than the old. Similarly, the new product may use more of one resource, less of another, and whether and how the overall resource use has changed depends on the system of weighting adopted. Nevertheless, as with the definition of technical efficiency of techniques, the distinction is useful in highlighting the effects of technical change on the technological choice before underdeveloped countries.

Product obsolescence due to greater efficiency of products occurs irrespective of income levels, and therefore applies as much to underdeveloped countries as to the advanced countries, whereas product obsolescence associated with higher incomes may not apply to societies with lower incomes. Suppose, following Lancaster's presentation, we postulate two characteristics possessed by products *j* and *k*, shown on the axes of Figure 1.7. Any particular product then represents a unique combination of the two characteristics, shown by a point of the diagram, say *A* below. Increasing quantities of the product involve increasing quantities of the two characteristics, in the same ratio as for *A*, as shown by the ray from the origin going through *A*.

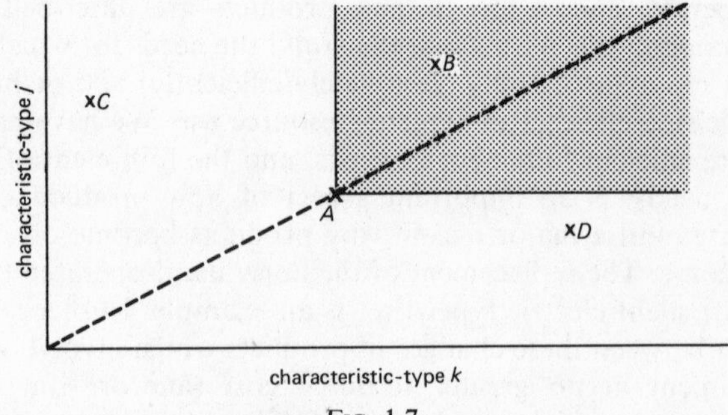


Fig. 1.7

Technical change in products consists in changing characteristics of products such that (i) for any given resource cost the total quantity of need fulfilment rises;[19] (ii) the type of need fulfilment changes, following changes in cultural and economic circumstances. In diagrammatic form the first type of change means that the same resources produce more of both types of characteristic (or more of one and no less of the other), and can be represented by the quadrant to the 'north-east' of *A*, shaded in the diagram. The second type of change involves a changing balance of characteristics, with the new products embodying more of one characteristic and less of another, as in a movement from *A* to *C* or *D*.

The first type of change makes product *A* inefficient or inferior, in a very similar way to the way in which techniques may become inferior owing to technical change. Such change therefore, as for techniques, applies to all societies irrespective of their economic situation because to adopt product *A* once *B* had been developed would involve a greater resource cost for the same need-fulfilment. The second type of change extends the range of products but does not make *A* inferior. In principle products *C* and *D* might coexist with *A*. However, economies of scale (not allowed for in the diagram) may mean that product *A* becomes a very costly product once market demand switches to *C* or *D*, and therefore *A* ceases to be produced.[20]

There is a third type of change not incorporated into the diagram. The diagram looks at the extent of need fulfilment with *given resource cost*. But products are indivisible, and each product not only represents a unique combination of characteristics but also a unique absolute quantity, and a corresponding unique absolute quantity of resources required to make it. In diagrammatic terms this means that different quantities of product are not represented by a continuous ray from the origin, but by a discontinuous line. Technical change may change the minimum absolute quantity of characteristics represented by each product, simultaneously changing the absolute resource cost represented by each product.

In practice, technical changes in products often embody all three types of change simultaneously: that is, the new products offer greater need fulfilment in relation to resource cost, change the balance of characteristics *and* embody greater quantities of resources. The last change is like increasing indivisibility with economies of scale in techniques. Technical change in products in developed countries occurs in line with the rising incomes (and is partly responsible for those rising incomes), so that the balance of characteristics offered by new products corresponds to the changing demands of consumers as their incomes rise. The rising incomes have the effect of shifting demand towards different products with more sophisticated, labour-saving, higher-quality, etc. characteristics: to summarise this complex of changes we may say that the characteristics of the new products have more high-income characteristics. Interpreting the two characteristics, *j* and *k*, as 'high-income', 'low-income' – see Helleiner (1975) – product changes in developed countries involve an increasing balance of high-income characteristics. If *P* is the initial product, changes for a given resource cost are in the shaded area in Figure 1.8 (for any movement below line *Pg* would involve a reduction in need-fulfilment in relation to resource cost, and any movement below *Of* would involve an increasing balance of low-income characteristics). Changes to the east of the vertical line from *P*, such as to *F*, make product *P* inferior. Changes to the north-west, such as *G*, leave *P* as an efficient alternative incorporating more low-income characteristics, but because of economies of scale in production *P* may become inferior. The third aspect discussed above, the increasing indivisibility of products, so that each product tends to represent[21] an increasing minimum quantity of need fulfilment and resources, can be thought of as itself a high-income characteristic, and is therefore represented by the changing balance of characteristics.

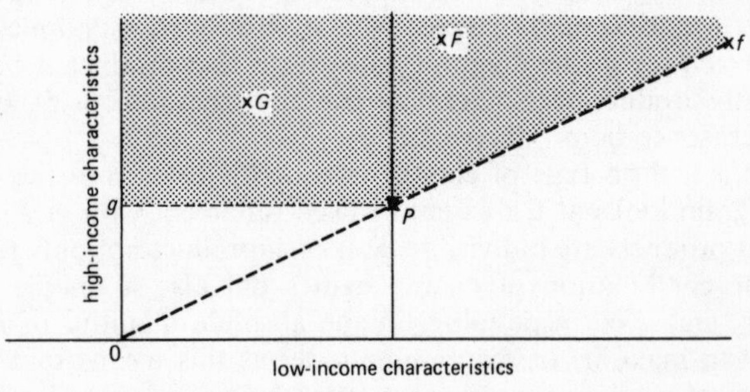


FIG. 1.8

As far as underdeveloped countries are concerned the products from developed countries tend to become increasingly high-income and therefore inappropriate for them over time. Product changes in the interests of underdeveloped countries would be those involving a greater balance of low-income characteristics, i.e. those *below* line *OPf*.

There is a strong parallel here between the discussion of technical change and obsolescence in methods of producing a given product, and of product development. For both there is one type of change which is dependent on the particular economic facts (relative prices and income levels) of the society in question; these changes are reversible. But there are other changes which make the previous techniques obsolete irrespective of economic facts, and these changes are not reversible without increasing the resource cost of production.

The range of efficient techniques to be included in the technological choice of underdeveloped countries, it would appear, should *exclude* those old techniques that have become technically inefficient, but *include* those techniques that have become obsolete because of changes in the economic structure in the developed countries, which have not been paralleled by similar changes in the underdeveloped countries. However, in practice these different types of change do not occur in different techniques so that they can be neatly separated: rather for the most part technical change of all types occurs simultaneously. That is to say on the factor-use side, changes have involved increases in efficiency and changed relative factor use, while, on the product side, the same product developments have been associated with increased income levels and reduced resource cost of need fulfilment. This is illustrated in Figure 1.9. Diagram (*a*) shows methods of producing the 'same'

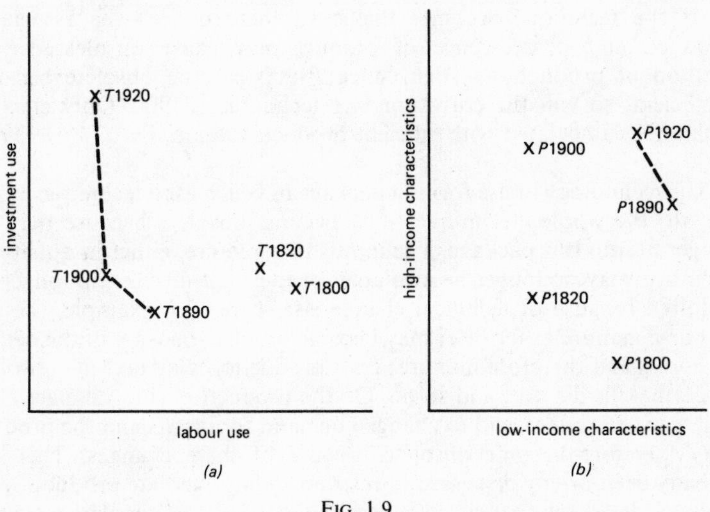


FIG. 1.9

product: the 1800 technique is initially replaced by a technique using more investment and less labour, but then there is a technical advance such that *T*1890 uses less labour and investment than *T*1820 and *T*1800, but more investment per man than the previous techniques reflecting the changing factor prices. Thus to benefit from the technical advances a country should choose *T*1890, but in doing so it will also have to accept the increased investment/labour ratio resulting from changes in factor availability. Similarly, there are changes on the product side (diagram (*b*)): assume that each of the products shown represents the combination of characteristics embodied in the products produced by the various techniques, so that *P*1800 is the product produced by *T*1800 and so on. To separate product efficiency and obsolescence from technique efficiency and obsolescence, we have to assume that each of the

*P*s in this part of the diagram is produced at the same resource cost – an incorrect assumption, which we relax immediately. *P*1820 involves more high-income characteristics, less low-income. A choice might be made between the two with lower-income countries/consumers selecting *P*1800. But *P*1890 is more efficient all round than either of the other products (because it embodies more of *both* characteristics), consequently the other two become obsolete. *P*1920 makes *P*1900 inefficient but not *P*1890. Hence from the two sides of the diagram it appears that the true technological choice – assuming inefficient techniques and products are rejected – consists of *T*1920, *T*1900 and *T*1890, and *P*1920 and *P*1890, and shown by the dotted lines. But the dichotomy between methods of production and products is a false one: each technique produces a particular product – each *T* corresponds to a particular *P*. If the technique becomes inefficient then so does the associated product, and obsolescence of product may cause obsolescence of method of production. Then, since *P*1900 became obsolete because inefficient, so will the corresponding technique, *T*1900. Only efficient techniques associated with efficient products remain, i.e. *T*(*P*)1890 and 1920.

The technology in use forms a package in which each technique has to fit into the whole; techniques may become obsolete because they no longer fit into this package. The inputs they require, either in quality or quantity, may no longer be available; or their output may no longer be required because of technical change elsewhere. For example, the use of horse manure as fertiliser may become obsolete because of the dearth of horses and therefore manure; craft production may become obsolete as craft skills die out, and so on. On the production side, changes elsewhere in the system lead to changed demand for inputs and the product may therefore become obsolete because of these changes. This has already been briefly discussed in relation to intermediate products, and in more detail for changes in general income levels. Complementarities between products both as inputs and in consumption[22] (eggs and bacon, horses and carts, cars and petrol, cars and roads, swimming baths and swimming costumes) mean that changes in one set of products change the market for others. The nature of consumption goods is also responsive to factor price changes with labour-saving devices being partly designed to save *hiring* labour[23] as well as representing an increase in income on the part of the consumer, by saving her labour. Administrative/managerial changes may also cause obsolescence: will the abacus survive large-scale methods of retailing? One could go on more or less indefinitely.

Much obsolescence due to changes in the technology in use can be classified, and has already been discussed, in terms of the earlier classification of changes in resource use and product demand. Equally, in one way all these changes could be interpreted as changes in the technology in use. However, it is worth having a separate category to bring out the importance of obsolescence due to changed technical requirements and technical inputs as opposed to economic changes – not changes in the availability of labour as a whole nor demand for luxury goods as a whole, but changes in the availability of steel of a particular quality at a particular price, or in the demand for a particular specification of blackboard chalk, that arise from the technology in use elsewhere in the system, changes in which often make a particular technique obsolete.

The important point is that technology is a package, and the efficiency or otherwise of a particular technique depends not only on its own performance, and that of its immediate substitutes, but also on the surrounding technology. This means that a technique may survive from the past as efficient (like *T*1890 and *T*1920) yet be obsolete in the context of the rest of the technology in use; selectivity may often be impossible because of linkages throughout the system.

To summarise: world technology consists of all those techniques that have evolved historically. But within this total many have become inefficient through scientific advances both in processes and in products. The remaining efficient techniques may none the less be obsolete because they form part of a technology package which is past. Such techniques would be efficient if the whole package could be revived, but this would involve reviving technically inefficient techniques, as well as efficient. The concept of package thus discredits the idea of assessing techniques individually, as efficient or not. The techniques which do survive reflect, in large part, the economic conditions when they were developed. Because of scientific and technical advances over time, and because of the package aspects, the more recently evolved they are the more likely it is that the techniques do represent an efficient alternative.

We have been primarily concerned with technology developed in the advanced countries. In addition traditional technology already in use in underdeveloped countries adds to the technological choice. Generally speaking, this technology makes little use of scientific and technological advances and is therefore often inefficient as compared with the advanced alternatives. Being developed over a long time in underdeveloped countries, it is likely to be suited to the factor availability and income levels in those countries where it was developed.

The real technological choice consists of that subset of efficient world techniques which is known to those making the choice, and which are available. Channels of communication therefore help determine the choice, with such channels, generally speaking, being more effective the more recently developed (within limits) the technique and the more it is being used and produced. Techniques that are obsolete in the developed countries may get lost in the mists of time and not present a real choice for that reason. The range and effectiveness of communication depends on who is doing the communicating and why: machinery producers will publicise their own products, while other institutions, like the Intermediate Technology Development Group, publicise a different type of technique. Apart from knowledge, there is also the question of availability. Second-hand machines extend the choice beyond those currently being produced, but with this exception, the choice is limited to techniques currently being produced, which for the most part means those being installed in the advanced countries, for those countries lacking facilities of production themselves. Again this further limits the choice toward techniques recently developed for use in the advanced countries.

### SELECTION MECHANISMS

It was argued earlier that each technique is associated with a vector of characteristics – consisting of the inputs required, quantitatively and qualitatively, the nature of the product, the scale of production, productivity of the various factors, the organisation to which it is best suited and so on. These characteristics tend to reflect – or at least be in tune with – the economic/historic circumstances of the economy where the technique was first introduced. Those who introduce techniques into underdeveloped countries thus make a choice among the techniques available; the choice actually made depends on the nature of the decision makers and their objectives, the economic circumstances in the economy concerned, and the characteristics associated with different techniques, bearing in mind that their choice is confined to the techniques they know about, and that their knowledge may often be incomplete or inaccurate.

Decision makers differ as to motive, knowledge and circumstances, so who takes the decision may determine what decision is made. For example, a subsidiary of a multinational firm may have, as prime motive, maximisation of profits after tax, on a worldwide basis. Locally and privately owned firms may aim to maximise local profit after tax. This difference can make a considerable difference to choice of technique in terms of nature of output, scale, specialisation, type of inputs used, price of such inputs, etc. A government-owned corporation may aim to maximise local profits before tax; it may also include other aims that are given little weight by the private sector – e.g. employment maximisation, or the spread of opportunities to the rural areas. The aims of those taking the decisions may differ from those of the corporations for whom they decide – individual income and/or prestige maximisation may alter decision making, sometimes allowing corruption to be decisive in choice of technique. The aims of family enterprises are likely to be in terms of total income of the enterprise, rather than profits.

The circumstances in which firms operate also differ as between the type of operator. For example, access to funds for investment, in quantity and quality, differs between firms. Multinational firms may obtain more or less unlimited funds at the cost of funds for the group as a whole. Local large-scale firms may borrow from the banks, often at interest rates which are held low to encourage investment. Smaller-scale enterprises, including family enterprises, may find it difficult to raise funds in any quantity and may have to pay high prices. Different types of firms tend to serve different markets: for many subsidiaries of multinational firms the world is their market. Locally owned firms tend to be more confined with the larger-scale firms serving the upper income groups, and doing some exporting. Family enterprises, particularly in the informal sector,[24] tend to produce for the consumption of those in the immediate vicinity, particularly among the lower income groups. Scale of operations is a function of organisation, availability of funds and the nature of the market. The scale of operations is often the decisive characteristic in determining selection of technique, with only one technique that is efficient at each scale. Another characteristic that is often decisive, as suggested by empirical studies,[25] is product specification. Product specification depends on the nature and income levels of consumers, and the structure of the economy as a whole. Different types of firms tend to have different product requirements – mainly because their markets differ, with the larger scale serving the higher income groups and competing on the world market,[26] while the small-scale and rural cater for local low-income consumers. Keeping up standards, the prestige of the firm generally and maintaining the value of the brand name also help determine product standards.

The price and availability of other inputs also differ between types of firm within any economy: it has been shown[27] that raw materials may be obtained at a lower price by the large-scale than the small. Firms with foreign technology contracts, and multinational firms, have access to inputs at different prices from those without. A major difference between types of enterprise is their access to different types of labour, and the price they pay for it. Factors holding up wages – such as government regulations, and trade union activity – are confined to the large-scale; family enterprises may generally obtain labour at a much lower price.

While there are major differences in objectives and circumstances between different types of firm within an economy, within each category selection of technique also depends on the way in which the economy as a whole operates. This is partly a question of price and availability of different inputs, including labour and investment goods; partly of income distribution determining the nature of markets; partly of the openness/closedness of the economy determining the extent to which products have to compete internationally. As argued above, the package aspect of technology means that any one technique which, looked at by itself, may appear efficient and appropriate, may be inefficient in the context of the technology in use. For example, a decision on the technique to be adopted in tyre manufacture will depend on the nature of the economy and income distribution within it – whether cars are being consumed locally, whether they are produced locally, or whether it is a bicycle economy, the standard of roads, the extent to which the tyres have to compete with other tyres manufactured locally or imported, the standard and prices of the competitive goods, the availability and price of inputs required, including energy, labour of different skills, materials and so on. The decision has to be made in the light of, and may be uniquely determined by, the nature of the economic structure as a whole. A system in which private firms compete, each with free access to foreign technology, may lead to oligopolistic competition via product differentiation – as it has in capitalist advanced countries – rather than competition via price. Such a structure may force each firm to adopt the most recent techniques in order to secure its market by providing the most recent product; in such a situation the technique is determined by the market in the context of the technology in use, although taken together the decisions also determine that technology. This is why it is difficult to induce marginal changes in technology. The implications, of this are further considered in the discussion of appropriate technology in Chapter 4.

To formalise, we may say that the process of selection of techniques consists of selecting from the matrix of known technically efficient and available techniques, which we described as *cT*, the nature of which was determined by the history of technological development, discussed above. Each technique within this matrix is represented by a vector describing its characteristics: $cT = (Ta, Tb, Tc \ldots Tn)$ where *Ta*, *Tb*, . . . are the different techniques, and each technique consists of a vector of characteristics *a*i, *a*ii, *a*iii, . . . associated with *Ta*; *b*i, *b*ii, *b*iii, . . . associated with *Tb*, and so on.

Decision makers may be categorised into groups, each of which has an objective function representing its aims. Suppose the decision makers are categorised into *M* (multinationals), *L* (large-scale local firms), *G* (government-owned enterprises), *F* (family enterprises) and so on. Corresponding to each group is an objective function, which we may represent as *m*, *l*, *g*, *f*. In trying to maximise their objective function, the decision makers are subject to a series of constraints, some of which are common to all of them, and some of which vary according to the category of decision maker. One such constraint is the nature of technology available – this may vary somewhat between decision makers, since knowledge about and access to different types of technology varies between groups – for example, a subsidiary of a multinational will have access to a different selection of techniques from a small family enterprise. In addition, as argued, markets, scale, factor availability and price may vary between the groups of decision makers. The underlying conditions in the economy, in contrast, tend to be common to all decision makers. We may describe the constraints of each group as *Cm*, *Cl*, and those common to the economy as a whole $C^*$. Selection of techniques then consists in the attempt by each group to maximise its objective function subject to the constraints. The overall balance of techniques within the economy depends on the size of the different groups. So $uT = f(m, l, g, f \ldots, Cm, Cl, Cf \ldots C^*)$, where $uT$ is that subset of total available techniques that are selected.

## THE NEO-CLASSICAL APPROACH

The neo-classical model of choice of technique[28] picks out two characteristics of techniques – labour and investment requirements – and regards the question of choice of technique as consisting of choosing between techniques of differing labour and investment intensity. The relative price of labour and investment is regarded as the determinant of this choice, with that technique being selected that maximises profits, given the relative price, and the substitutability between labour and capital. The approach, at its textbook simplest, may be shown with a smooth, convex isoquant representing different methods (in terms of $I/L$ ratios) which may be adopted to produce the 'same' output. In the figure below $R$ represents the profit-maximising equilibrium point, so that technique would be selected.

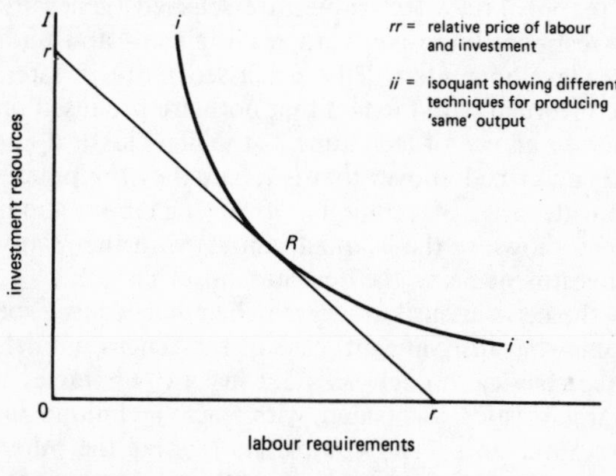


FIG. 1.10

Technical progress may be introduced without altering the basic model.[29] Technical progress is assumed to be 'neutral'[30] affecting all techniques equally, Thus the entire isoquant is shifted inwards over time, so that at any one time choice of technique is between techniques as represented by an isoquant even though the isoquant is shifting over time. The figures below illustrate how choice of technique is affected by technical progress.

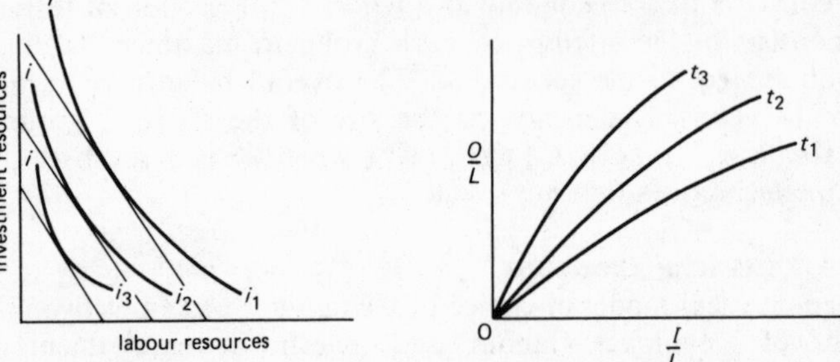


$ii_1$ is the isoquant in period 1 corresponding to the production function $OT_1$; with technical progress $ii_1$ shifts to $ii_2$, and $Ot_1$ to $Ot_2$, and so on.

FIG. 1.11

Developments of the neo-classical approach in application to underdeveloped countries have concentrated on two aspects: first, that savings generated per unit of investment should be a (sometimes *the*) criterion for choice of technique;[31] secondly, that the ruling relative price of labour and investment may be 'distorted', with the consequence that socially inappropriate techniques are selected (generally speaking excessively investment-intensive) with resulting un- and underemployment.[32] These developments will be discussed more in later chapters. Here we are concerned with the fact that both are premised on the same basic approach to choice of technique as the neo-classical approach, at its simplest, as described above: that is to say they are premised on the existence of a wide range of techniques of varying labour and investment requirements as shown in the isoquant above, with the relative price of labour and investment being the determinant of choice.

In one way the neo-classical model may be regarded as a special (and, in practice somewhat insignificant) case of the general model described above. The neo-classical model picks out just two characteristics of the manifold characteristics associated with each technique – investment and labour requirements – and completely ignores the others, such as scale of output, nature of product, skilled labour requirements, material inputs, infrastructural requirements etc., etc. In terms of selection mechanisms it concentrates on just one – relative prices of labour and investment – corresponding to *one* type of decision maker, the profit-maximising entrepreneur with unlimited access to finance at constant rates of interest. In order to achieve this simplification the model makes a *ceteris paribus* assumption about the many other influences over decision making, and about characteristics of techniques that determine technical choice.

It might be argued, none the less, despite the overwhelming empirical evidence that labour and investment intensity are not the most significant variables nor their relative price the sole critical determinant of choice, that the approach highlights an important subsection of technical choice – that concerned with the investment/labour intensity aspect of producing given output. Even this much cannot be granted because of the nature of technological development over time, which makes nonsense both of the assumption of a range of techniques of varying labour and investment intensity, and most significantly, of the *ceteris paribus* assumption about other characteristics of techniques.

As argued above, techniques have to be developed historically and the techniques available for selection at any one time are those that have been developed at some time in the past. Generally speaking the techniques developed at one time reflect the resources of the economy when developed. In terms of investment and labour intensity this means that more labour-intensive techniques are likely to have been developed at earlier periods, when savings were lower in relation to labour supply, while more investment-intensive techniques were developed later when more savings were available in relation to labour supply. In so far then as a neo-classical type range of techniques is available (i.e. techniques of varying $I/L$ ratios), it consists of techniques developed over a historical period with the more labour-intensive techniques dating back to an earlier period than the investment intensive. (This was illustrated in Figure 1.9 (*a*) above). But because the earlier techniques originate at an earlier time, they have less scientific and technical knowledge to back them up, and therefore tend to be of lower productivity, Many of them have become technically inefficient, as argued above, using more investment as well as labour in relation to output produced. Thus, far from there being a complete isoquant corresponding to each moment of time, for each scientific and technical age, as depicted in Figure 1.11, there is a series of techniques developed at different times, with a tendency for the earlier ones to become technically inefficient, as shown in the earlier Figures 1.6 and 1.9 (*a*).

In some ways this view of technical development is similar to that of Salter, (1966), and of Atkinson and Stiglitz (1969). Salter believed, with the neo-classicists, that potentially there was a whole range of techniques of varying investment and labour intensity, corresponding to each level of scientific and technical development, or 'state of technique' as he puts it. But of this large potential range, only the immediately profitable were actually developed; he contrasted the 'relatively narrow range of developed techniques actually available to businessmen – the range exemplified by the machines on the market', with 'the much wider range of techniques which could be designed with the current state of knowledge'.[33] The former – actual machines – are confined to a narrow range of investment and labour intensity reflecting the resources when they were developed. They might therefore be represented by a point on a diagram, as in the Figures above (e.g. 1.6). Atkinson and Stiglitz discuss the localised nature of part technical progress, consisting of improvements in techniques already in use, as shown in the Figure below.

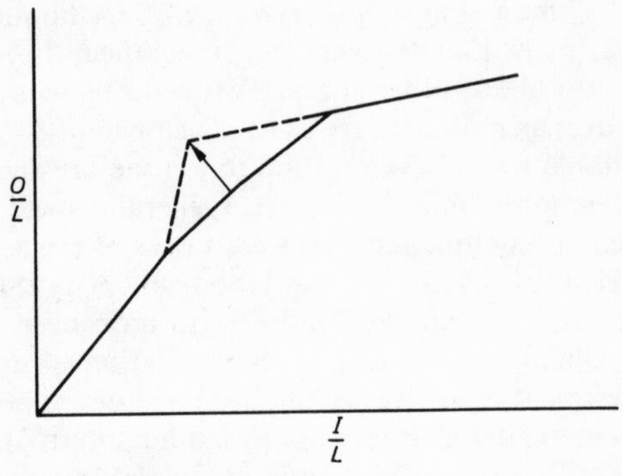


FIG. 1.12 Atkinson/Stiglitz localised technical progress

Both Salter and Atkinson and Stiglitz stick to the neo-classical framework in two respects: first, showing a production function, whether potentially (Salter) or as the starting point (Atkinson and Stiglitz), despite the fact that the conclusion of a historical approach to technical development must be that there is little, if any, meaning to the assumption of an almost continuous range of techniques reflecting possibilities at a single 'state of knowledge'; all there is are *developed* techniques which have been developed at different times. Whether or not it makes sense to talk of a potential production function, along with Salter, is really a matter of metaphysics. More relevant for our discussion, is their concentration, as in the basic neo-classical approach, on the two neo-classical characteristics of techniques of labour and capital intensity, ignoring the other characteristics, and selection mechanisms. This is significant because these other characteristics are systematically related to each other and to the investment intensity of techniques, and tend therefore to invalidate any conclusions which ignore them. One aspect discussed at length earlier, is product development which occurs systematically over time, and which means that the later techniques are associated with different – more efficient and higher-income – products than the earlier techniques. Hence to assume that techniques of varying labour and capital intensity exist which produce the *same* product does not make sense, because the technical developments that have increased investment have also been associated with changed products. This also means, as argued earlier, that some early techniques which might appear to offer a labour-intensive alternative, are ruled out because associated with obsolete products. In a way this conclusion is ironic because many neo-classicists[34] have used product choice to rescue the idea of technical choice in terms of investment intensity – arguing that while empirical evidence supports the idea of relative coefficient fixity for each product, product choice allows coefficient variability. Other systematic changes in characteristics of techniques over time are scale changes, and input (and particularly skill) requirements. Again this means that later techniques are designed for larger scale than earlier so it rarely makes sense to think of them as producing on the same isoquant. Changes in the skill availability and requirements, and other changes in the technology in use, also mean that later techniques are associated systematically with different characteristics than earlier ones. Since the only genuine choice of technique along neo-classical lines is the result of the survival of earlier labour-intensive techniques along with later more investment-intensive techniques, any choice between them also involves a choice of the other characteristics that have changed systematically with time. So the early techniques are designed, on the whole, for production at smaller scale, use less skills, require less technologically advanced inputs, and so on and so forth, as compared with the later techniques. In view of this it is scarcely surprising that investment intensity often becomes of subsidiary relevance to the choice, and the relative price of labour and capital also only of minor importance, compared with the other characteristics that are ignored in the neo-classical approach.

## CONCLUSIONS

This chapter has attempted to show how the historical evolution of technology has conditioned the characteristics of the technological choice facing underdeveloped countries. It has also discussed the factors determining the selection of techniques in any country, arguing that different groups of decision makers face different constraints, have different objectives and therefore make different choices. Finally, it is argued that the neo-classical model picks out only one minor aspect of the choice; takes an unhistoric view of the choice available; assumes a single type of decision maker; and in its *ceteris paribus* assumption ignores many of the most significant factors affecting choice, as the empirical section of this book shows. This chapter has been concerned to discuss the general factors affecting technological choice in underdeveloped countries; Chapters 3 and 4 look more specifically at the implications for underdeveloped countries' patterns of development of the technological choice and at the characteristics of an 'appropriate technology'. But first, in the next chapter we consider the nature of employment problems in underdeveloped countries and their relationship to the question of technology.

NOTES

1. R. S. Merrill (1968).
2. See e.g. Bernal (1969) for a historical description of scientific developments over time.
3. See Marglin's (1974) paper with the graphic title: 'What do Bosses do? The Origins and Functions of Hierarchy in Capitalist Production.'
4. An analysis of research and development expenditure in the U.S. in 1953 showed that large-scale corporations (with more than 5000 employees) were responsible for nearly 70 per cent of expenditure on research and development, medium firms (1000–5000) for 18½ per cent and firms with less than 1000 employees for 12½ per cent. See Schmookler (1966), Table 3.
5. According to Berger *et al.* (1973) packages may be *intrinsic*, when the individual items are inseparable, or *extrinsic*, when they *may* be acquired separately. The distinction is not analytic; items which are inseparable analytically are better described as a single item. Rather it is a matter of practical possibilities, cost and consequences of separation. It therefore depends on how the package was evolved historically and the alternatives that have been developed. This means that (i) there is no hard and fast line dividing intrinsic from extrinsic packages, rather it is a matter of judgement: and (ii) the same package may be intrinsic at one time and become extrinsic as new possibilities develop, as with the gradual increase of alternatives to the package represented by multinational investment. Where *package* is used in the above discussion it is normally taken to be intrinsic, judged by current possibilities, cost and efficiency; new developments would be required to make it extrinsic.
6. Doyle's study (1965) of two cement plants provides an example of how a technique may require specific inputs of infrastructural services; this led to considerable investment in Indonesia, where they were not previously available, as compared with the U.S. where they were.
7. The concept used by Myrdal in *Asian Drama* to encompass all items like nutrition, health and hygiene, relevant to the efficiency and well being of the people.
8. One study found that for 90 per cent of manufacturing firms the main *aim* of their research programmes was the development of new products or the improvement of old (see Gustafson (1962)). Other studies have found that three-quarters or more of research results involved substantial product changes – see Wagner (1968) and Bloom (1975). Even where product changes are not the main aim, they may none the less result from changes in techniques.
9. Much of the incentive for technical change has been attributed to imbalances – technical and economic – resulting from one technical change leading to a new supply/demand situation for other inputs, and hence to further innovation. This view has been fully developed by Rosenberg (1963(b) and 1969) with supporting empirical material.
10. Some (e.g. Wilkinson (1973)) have attributed much of the industrial revolution to the rising price of wood.
11. See the empirical section of this book.
12. See Adam Smith (1776) and Allyn Young (1928).
13. See e.g. evidence of Sands (1961), Armstrong and Silberston (1965), and Sawyer (1971).
14. According to this definition a technique is technically inefficient as compared with another technique if it uses more of one input and no less of another to produce the same output. The definition must be distinguished from engineers' and physicists' definitions of technical efficiency, which are concerned with the physical properties of the technology – e.g. extraction rates, power loss. It must also be distinguished from *economic efficiency*, which is related to cost of production. Methods which are technically inefficient in engineering terms may none the less be technically efficient as defined here because reduced output rate is

more than offset by reduced inputs. Methods which are economically efficient must also be technically efficient, as defined here, so long as factor prices are positive.

15. The thorny question of measurement of investment and investment intensity is dealt with in Chapter 8, as is the relationship of this question to that of measurement of capital. For most of this discussion above we assume that it is possible to arrive at some measure of investment.
16. See Galbraith's (1967) account of how this operates in advanced countries.
17. Sercovich (1974) describes this process in Argentina.
18. This is the approach to products and consumer demand adopted by Lancaster (1966).
19. Technical change in methods of production also produces this result but through changes in the method of production, not changes in the nature of the product. This is a theoretical distinction. In practice, as argued later, technical change occurs simultaneously in products and methods of production.
20. See Lancaster (1975).
21. This simplified representation describes how new products, once established, compare with old. It does not describe the different stages of development of a particular product (as e.g. Vernon (1966)).
22. See Streeten for a discussion of complementarities between consumption goods.
23. Indicated in the Edwardian statement: 'the best labour-saving devices are servants'.
24. Defined in Chapter 2.
25. See Chapters 8 to 10 below.
26. See Chapter 7.
27. E.g. by Prasad (1963).
28. Sen's classic work (3rd ed. 1968) provides a good example.
29. See e.g. Meade (1961).
30. The various definitions of neutrality (e.g. Hicks, Harrod, Solow) need not concern us here.
31. See e.g. Galenson and Leibenstein (1965), and Sen (1968).
32. See e.g. Bruton (1974).
33. Salter (1966) pp. 14–15.
34. E.g. Solow (1970).

# 2 The Employment Problem – a Conceptual Discussion

Employment, or rather its inadequacy, is a problem of major concern in most discussion of third world issues.[1] Advanced technology from developed countries is viewed with suspicion largely because of its failure to provide sufficient jobs. Yet there is little agreement about how the employment problem should be defined, or measured. While there is consensus that there is a major, possibly overriding problem, of an unspecified nature, when it comes to specification – and to an even greater extent proposed cures – concepts overlap, conflict and confuse. Analysis of the employment question is peculiarly dependent on analysis of technology questions, since technologies are largely categorised in terms of their employment effects. This chapter discusses some of the conceptual differences in approaches towards employment, and considers briefly how the different approaches give rise to alternative approaches towards technology.

Difficulties arise, as so often in development economics, from the attempt to apply concepts developed in and for developed countries to underdeveloped countries.[2] Both *employment* and *unemployment* are concepts related to economies where wage-labour predominates. Employment normally means working *for* an employer for wages, while unemployment is defined as the active seeking of employment, at the ruling wage-rate. Wage-labour, as a mode of production, develops with industrialisation. In pre-industrial revolution societies other forms of labour relationship predominated – feudal ties of one kind or another, slavery, family and self-employment. In Britain in the eighteenth and nineteenth centuries, the enclosure movement in agriculture, and the factory system in industrial production, substituted wage-labour for traditional forms of work. Underdeveloped countries are in this respect in a transitional state. The modern industrial sector, generally initiated by expatriates from industrialised societies, uses the technology, managerial techniques, and labour mode of the industrialised countries, and relies for the most part on wage-labour.[3] But in the rest of the economy – still in a pre-industrial state – other traditional forms of labour relationship are prevalent. Wage-labour may occur in a sporadic, seasonal way supplementing incomes and activity, but is by no means the main source of either. Of course, in this respect, as in so many others, developing countries differ enormously.[4] The form of work that predominates depends on many factors – the stage of development and the extent to which the market has penetrated the subsistence sector, forms of land tenure, land/labour relationships, and other historical factors. But it can be fairly concluded that proletarianisation, or the predominance of the wage-labour relationship, occurs with industrialisation.

Where there is wage-labour, employment may be precisely defined: it occurs when a person is working for an employer for wages. Unemployment is defined as occurring when a person is without employment and

TABLE 2.1 Numbers in wage and salary employment as %

| *Country* | *Economically active* | *Population of working age (15–60)* |
|---|---|---|
| *Underdeveloped* | | |
| Algeria | 60·2 | 27·5 |
| Egypt | 53·9 | 29·2 |
| Maroc. | 37·3 | 19·9 |
| Sierra Leone | 10·9 | 8·7 |
| Tunisia | 61·7 | 29·7 |
| Argentina | 70·8 | 42·5 |
| Brazil | 54·8 | 31·8 |
| Colombia | 57·3 | 33·4 |
| Chile | 70·1 | 36·7 |
| El Salvador | 48·4 | 35·6 |
| Mexico | 62·2 | 33·3 |
| Uraguay | 69·8 | 42·1 |
| India | 17·0 | 10·8 |
| Indonesia | 31·7 | 20·1 |
| Iran | 43·8 | 26·5 |
| Korea (rep.) | 38·0 | 23·0 |
| Philippines | 39·9 | 25·1 |
| Singapore | 68·5 | 41·5 |
| Sri Lanka | 54·3 | 33·5 |
| Thailand | 15·4 | 14·6 |
| *Developed* | | |
| France | 79·0 | 55·7 |
| Germany | 85·1 | 59·7 |
| Sweden | 92·0 | 62·9 |
| U.S. | 90·3 | 65·7 |
| U.K. | 90·1 | 64·4 |

SOURCE *Annual Yearbook of Labour Statistics, 1974* (Geneva: I.L.O., 1974) Tables 1 and 2A.

is actively seeking a job as an employee. Even in developed countries the two concepts by no means cover all the relevant categories. Much productive economic activity occurs outside the formal employment situation. Where such activity results in marketed output, those involved are said to be self-employed. Where the activity is home production for home consumption, like the work of married women in the home, do-it-yourself building, decorating, etc., the activity is excluded from employment and output statistics. As the figures in Table 2.1 show, large numbers are omitted from both employment and unemployment in the statistics for developed countries, although a high proportion of the 'economically active' population are in wage and salary employment (80 per cent or over). In underdeveloped countries much less than half of the population of working age are in wage and salary employment.

Moreover, the concepts are designed to cover individuals who fall exclusively into one category – be it employment, unemployment or self-employment. Part-time employment which may be combined with voluntary idleness, with activity in the home, with self-employment or with the search for additional work, is not adequately covered.

The concepts, therefore, do not do justice to the manifold productive activities in developed countries – they are based on an image of society in which everyone who is not actually earning money (described emotively as 'gainfully' employed) is sprawled out in front of the television, as near to complete idleness as one can get without actually being unconscious. Not only is this a travesty of reality, particularly that of women's work in the home, it also encourages people to adapt to the image, becoming psychologically as well as economically dependent on wage-employment. None the less, it is true in developed countries that employment provides the majority with the main means of livelihood, and that, in the absence of state subventions lack of employment is a major source of poverty. But in most third world countries, employment in the modern sector is a source of livelihood for only a minority of the population. Many more work in traditional modes, working directly for themselves in subsistence activities, self-employed on marketed activities, and working for others on a traditional rather than wage-employed basis, such as share cropping. Traditional forms of activity are not confined to agriculture; traditional crafts are practised on this basis. Services too, extending from traditional judicial functions, medicine, entertainment, to moneylending, are also largely outside the wage-employment nexus. With growing urbanisation, following industrialisation, new forms of urban activity have developed – the so-called 'informal' sector – in which people perform a variety of functions, manufacturing, services and construction for themselves and for the modern sector in and around rapidly growing towns.[5] There are some employees attached to the most successful of those in the informal sector – though they rarely appear in the statistics because of the informal nature of the sector. But to a great extent the informal sector is organised on a self-employment basis. Many of those who are employed in LDCs are employed on a part-time basis, supplementing their incomes from other activities, and not relying on employment for their exclusive means of support. The application of the advanced country concepts to third world countries leads to strange conclusions: employment then accounts for only a minority of those of working age. Should the rest therefore (in some countries accounting for over 90 per cent of those of working age) count as unemployed? Statistics for rates of open unemployment would suggest not. In many countries rates of unemployment are not all that high.

There are three reasons why unemployment rates are so low in relation to the large numbers who are not employed in the modern sector and who might therefore be thought of as potential *unemployees*. One is that although they are not in employment, as defined in advanced countries, they are at work, producing means of livelihood for themselves and their dependants. Secondly, unemployment is a luxury – a luxury for (relatively) rich individuals in very poor societies, whose families must somehow support them while they are looking for work;[6]

TABLE 2.2 Rates of unemployment in selected countries, 1973

| *Country* | *Unemployed* ('000) | *As % wage/ salary earners* | *As % population of working age* |
|---|---|---|---|
| Ghana | 26·3 | n.a. | 0·6 |
| Egypt (1972) | 134·6 | 3·0 | 0·9 |
| Maroc | 29·1 | 2·0 | 0·3 |
| Sierra Leone | 5·9 | 5·8 | 0·3 |
| Tunisia | 37·0 | 5·5 | 0·2 |
| Zambia | 9·3 | n.a. | 0·4 |
| Argentina | 174·2 | 2·7 | 1·2 |
| Chile | 93·1 | 5·1 | 1·9 |
| Guatamala | 0·6 | 0·08 | 0·02 |
| Peru | 199·8 | 12·9 | n.a. |
| Puerto Rico | 112·0 | 15·1 | 7·3 |
| Uraguay | 49·4 | 7·0 | 2·9 |
| India | 7713·8 | 25·1 | 2·7 |
| Indonesia | 84·3 | 0·7 | 0·1 |
| Korea (Rep.) | 461·0 | 11·7 | 2·7 |
| Pakistan | 167·6 | 5·3 | n.a. |
| Philippines | 624·0 | 12·7 | 3·2 |
| Singapore | 35·7 | 7·2 | 3·0 |
| Sri Lanka | 457·7 | 19·1 | 6·4 |

SOURCE *Annual Year Book of Labour Statistics, 1974*, Tables 1, 2A and 10.

and a luxury for rich societies which can, as nearly every developed country does, afford to provide for the unemployed as a class, via a system of unemployment benefits. Thirdly, where wage-labour is a minority activity most people still have some connection with family enterprises, so that people may *choose* whether to be unemployed – that is actively to seek for work – or to work in the family activity. To a much greater extent (though not entirely) in developed countries, there is no choice. Unemployment is the only alternative to employment.

For these reasons unemployment is a very different phenomenon in underdeveloped countries than it is in developed countries. In developed countries, it includes all those for whom the system simply fails to provide work, whereas in underdeveloped countries many of the unemployed have chosen to be unemployed. To the extent that they have so chosen, they are often not the worst off in society. Those in the rejected activities who have neither means to support unemployment, nor hope of finding employment, may be worse off.[7] Analysis of the unemployment figures in underdeveloped countries, therefore, must concentrate on sorting out the factors leading people to make this choice, rather than accounting for a total deficiency of work opportunities as in Keynesian-type analysis for developed countries. Todaro's work is an example of this type of analysis. His model suggests that unemployment is chosen on the basis of two factors: the income differential between employment in the modern sector and work in the agricultural sector, and the estimated probability of finding employment in the modern sector. While there are many criticisms of the details of the model, the basic approach correctly shows open unemployment in third world countries to be of a very different type from that in the industrial countries. In the latter, it may be viewed as *the* crisis of the system, since employment represents, for the most part, the only means of gaining a livelihood. But for underdeveloped countries, unemployment is to a large extent *chosen*, and is the consequence of major differentials in rewards between the modern industrial and traditional sectors. Unemployment is created by industrialisation: without a modern sector there could be no one seeking a job in the sector.

Yet the fact that open unemployment is not overwhelmingly large does not mean that the distribution of work activity as a whole is satisfactory. Many have seized on concepts of 'disguised unemployment'[8] and 'underemployment' to describe those who are not actively seeking work, but whose activities are in one way or another unsatisfactory. The point is to redefine the advanced-country concept of unemployment to cover those whose situation is believed to be akin to that of the unemployed, despite apparent activity. Some have been tempted by Western concepts to believe *all* those outside the modern industrial sector to be potential labour supply for that sector. Others have been more discriminating.

There are two distinct aspects to the concept of open employment. One is the volitional/frustration element: that is that the openly unemployed are identified only if they are actively seeking work. The other is the surplus labour aspect: that is, the unemployed form a pool of extra labour which the industrial economy may absorb. The point about the openly unemployed is that both factors must be present. Others, e.g. married women at home, may form a potential pool of additional labour, but they do not count if they are not actually looking for work. Equally, there may be people who are frustrated in their jobs, overqualified and unsuited, and looking for promotion or a shift in employment, but they do not count as openly unemployed, and do not form a potential additional labour supply, since they are already employed. The openly unemployed form a small overlapping segment of both groups.[9]

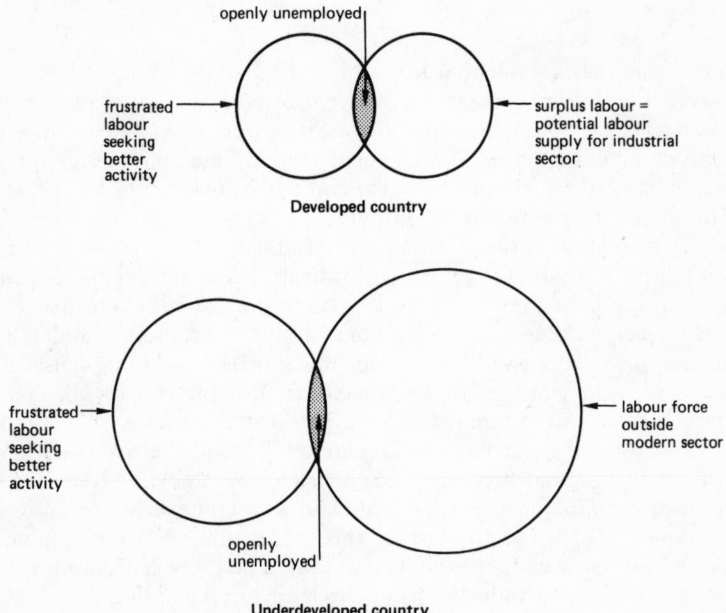


FIG. 2.1

In underdeveloped countries, the openly unemployed form an even smaller proportion of both groups. On the one hand the frustrated employees tend to be larger in numbers – because few can afford to be unemployed, and because the labour force tends to be overeducated in relation to job requirements, as education becomes a means of rationing the much sought-after higher-paid and higher-status activities.[10] On the other, the traditional sector is thought to form a pool of potential surplus labour. It is therefore tempting to extend the concept of unemployment to cover these elements, in particular the surplus labour/labour supply for the modern sector aspect. This is what the concepts of *underemployment*, *disguised unemployment*, etc., are intended to achieve. But this has led to difficulties. The openly unemployed are more or less visible and easy to count. The potential labour supply/surplus labour elements are not. They consist of the majority of the population working in a vast range of activities with varying productivity, organisation, hours of work, skills, education, willingness and ability to work. Most attempts to define and measure underemployment falter on the multi-dimensional nature of the activities of those who are not in modern sector wage employment. Above all they come up against the fact that many of the people concerned are working, often long hours,[11] so that to define them as akin to unemployed, does seem somewhat arbitrary and paradoxical.

### MARGINAL PRODUCT APPROACHES

Nurkse put forward the most straightforward definition of surplus labour: 'These countries suffer from large-scale disguised unemployment, in the sense that even with unchanged techniques of agriculture, a large part of the population engaged in agriculture could be removed without affecting agricultural output.'[12]

Empirical studies, however, have established, not really surprisingly, that marginal product is generally positive.[13] Positive marginal product per working hour may none the less be reconciled with zero marginal product per worker, if workers are working short hours, and if the remaining workers would make up any shortfall in hours caused by removing some portion of the workforce. But for this to take place automatically and voluntarily, if the workforce operates on marginal principles, there must be a particular relationship between marginal utility and marginal product, as Sen (1966) has shown. Alternatively, minor or even major organisational changes might ensue, preventing a loss in output. Nurkse allowed for this: 'One, thing, however, we need not and probably cannot exclude and that is better organisation. If the surplus labour is withdrawn from the land the remaining people will not go on working in quite the same way. We may have to allow for changes in the manner and organisation of work, including possibly a consolidation of scattered strips and plots of land.'[14]

The apparently straightforward measure thus turns out to be considerably more complex and the results depend on assumptions made about policy changes, worker motivation, what is to count as 'normal' work hours and so on. Nurkse was concerned with agriculture but a similar definition and discussion would extend to those in the urban informal sector.

One question that arises is why marginal product need be *zero* for underemployment to arise. True, only if it is zero will the release of workers cause no loss in output, but, with very low marginal product relative to the new occupation, the release of workers may still increase overall output substantially. Accordingly, workers with low productivity relative to the modern sector do form a potential labour supply for the high productivity sector. Sabot's definition of underemployment as occurring where marginal product per worker differs between sectors provides a measure of potential labour supply on this basis. Underemployment then occurs in any situation where net output could be increased by reallocation of the workforce. An alternative description would be Pareto-inefficient employment. Such an approach is, of course, still subject to the general difficulties of measuring marginal product – indeed in a way greater difficulties than the simple Nurkse approach because it involves measuring marginal product in the modern sector as well as elsewhere. Sabot solves the problem by assuming that marginal product is equated with the wage – a dangerous assumption, in the modern sector because of economies of scale, and elsewhere because of the nature of family employment and rewards. This approach to the measurement of underemployment would include substantial underemployment in almost all economies including developed countries, where wages and productivity differ between sectors. The definition makes the underemployed akin to the openly employed in the sense that output would increase were they re-employed, but in other respects they are not unemployed. They do not fully fall into either of the two circles described earlier.

Fei and Ranis (1975) define surplus labour as occurring where the wage *exceeds* the marginal product. The Sabot approach which assumes equation of wage and marginal product thus denies the existence of such underemployment. Presumably the justification of the Fei/Ranis definition is that where the wage exceeds marginal product, wages or incomes are performing some sort of relief function indicating that marginal product of labour is too low for subsistence. Since the additional workers are, therefore, consuming more than they are producing, they might be described as surplus to the sector. As before, the Fei/Ranis measure requires estimation of marginal product per man which again is dependent on the assumptions made about the response of other workers and subsequent organisational changes.

None of the suggested definitions are satisfactory because of these measurement problems; nor do they really identify a state akin to unemployment in the traditional sector because many of the workers concerned may be working long hours. The problem identified is not absence of work, but the nature of work. If the aim is to estimate potential labour supply for the modern sector, they succeed in a more or less rough way. But these are oversophisticated tools to achieve this.

From one point of view, irrespective of exact marginal productivity, most of those in the traditional sector form such a potential supply. From another point of view, as Myrdal emphasised, few may, because there are supply constraints – cultural, institutional and nutritional – which limit the number who would actually be suitable and available for modern sector jobs. These limits are unrelated to their current marginal productivity in the traditional sector. If the aim is to identify the potential labour supply, these institutional variables are the relevant ones to examine, rather than marginal productivity. In any case, for many economies the growth of modern sector employment is likely to be considerably less than potential supply, as is shown at the end of the chapter. Thus measures of potential labour supply to the modern sector are not particularly useful. What is needed is an examination of trends in employment and productivity in each sector.

### POVERTY APPROACHES

Those who are underemployed, according to any of the definitions discussed, are also poor. Recently, there has been a tendency to abandon the somewhat sterile exercise of measuring marginal product, and to define the underemployment problem as one of poverty. According to the I.L.O. Mission to Colombia: 'Poverty therefore emerges as the most compelling aspect of the whole employment problem in Colombia'. And an I.L.O. group of experts, meeting in 1966, concluded: 'Underemployment exists when a person's employment is inadequate, in quantitative or qualitative terms, in relation to specified norms'.[15] This includes those whose 'earnings are abnormally low'. Turnham came to a similar conclusion: on the 'assumption that some work is always available in the traditional sector and that numbers can be accommodated there partly through accepting lower income for a given effort', part of any excess supply of labour will be absorbed in the traditional sector, lowering productivity and incomes. Consequently, a survey of poverty would reveal the extent of underemployment.

Despite the fact that the term 'underemployment' is for the most part retained, the poverty approaches in effect abandon any relationships with the original concept of unemployment. Low average incomes are the defining characteristic of a poor society. Thus to identify underemployment with poverty comes very near to equating underemployment and underdevelopment. But these approaches define underemployment in terms of *relative poverty*, not absolute poverty. Turnham, for example, suggests including all full-time workers whose incomes fall 'below some reference level – say one third or less of average full-time earnings of the employed population'. This then makes the extent of underemployment dependent on the income distribution in the society. In a completely egalitarian society, however poor, there would be no underemployment; yet if, in such a society, a small proportion of the workforce acquired highly paid jobs, underemployment would emerge, even though the actual activities of the rest of the workforce remained exactly as in the previous situation.

Measuring the extent of underemployment by the extent of relative poverty gives results which are both arbitrary and artificial. Arbitrary, because where the line is drawn – one-third or one-fifth, etc. – is essentially arbitrary. Artificial, because the problem that has been identified – relatively low-income work – bears no relationship to the original concept of unemployment. Analytic (and policy) confusion is created by the use of the term, with the implicit assumption that the cure for the problem, as the cure for Western unemployment, is the creation of jobs. The obvious cure for low productivity is not the creation of more jobs, but improvement in productivity. The argument has been put forward forcibly by Weeks:

> . . . once it is conceded that the problem is income, or the inadequacy of income, we have granted that we are dealing with what has been called in the United States the 'working poor'. Of what use is it to call those who toil (for inadequate incomes) 'unemployed'? I suggest that the reason for this is political . . . If the problem is 'unemployment' then the solution is 'employment'. And if by employment we mean 'wage employment', then the solution is clear – faster growth of the foreign-owned 'modern' (i.e. imported from the West) sector. This is not the solution. It is the problem itself.[16]

The obvious next step after the poverty approach to measurement of underemployment is the step taken by Weeks: rejection of the concept of underemployment ('This . . . renders "unemployment" analytically redundant'), and concentration on poverty. But this too raises questions: why poverty, why not riches as the focus of discussion? Where should the line be drawn defining the 'working poor'? The two questions are related. To understand the source of poverty, one has also to examine the source of (relative) riches: to define the 'working poor' and analyse their position in society it makes no sense to examine them in isolation from the rest of society. Poverty can only be defined in a social context. Open unemployment too arises from the workings of the whole economy. It is not an isolated phenomenon that makes sense to examine on its own. What this means is that description and analysis should start with what people are doing or not doing: analysis of employment throughout the economy should precede analysis of unemployment and sources of poverty.

Let us illustrate the discussion with a schematic representation of how people are occupied in an underdeveloped economy. Figures 2.2 and 2.3 represent the model.

$LL^1$ in both diagrams represents the population of working age.[17] Figure 2.2 shows a pre-industrial society before the introduction of

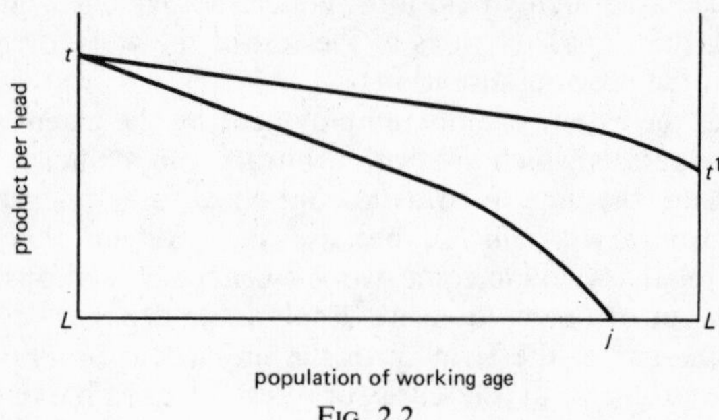


FIG. 2.2

Western technology. $tt^1$ shows how average product (including all subsistence products) varies with the size of the working population. It is assumed that, as numbers increase, there is a tendency for average product to fall because of the increase in the labour/land ratio. In a society with no industrial sector there is no open unemployment – since there is no wage-employment in the industrial sector to search for. All are occupied in one way or another in the traditional sector. Average product is then $L^1t^1$. If the product is equally distributed everyone receives $L^1t^1$ and there is no relative poverty. Thus neither the concept of open unemployment nor that of relative poverty makes sense. The Nurkse definition of zero marginal product would make sense assuming that marginal product falls to zero. The diagram has been drawn to include $jL^1$ of this type of underemployment where $tj$ is the marginal product curve, although, as noted earlier, identification of marginal product raises severe conceptual and measurement difficulties. The only other definition of underemployment that makes sense is the Fei/Ranis definition; this would extend to the whole sector since all are assumed to receive their average product, which is in excess of the marginal product. The Fei/Ranis definition, like the relative poverty definition, is critically dependent on how income is distributed. If all were paid their marginal product there would be no Fei/Ranis under-employment – but then there would be a problem of how to distribute the surplus. In an economy with no rentiers, and no employers, the whole product has somehow to be distributed to the working population, and therefore payment according to marginal product only makes sense with constant returns to employment, when marginal and average product are equal. This is one reason why a marginal product distribution system does not make sense in sectors where the employer/employee relationship does not prevail. If income is unevenly distributed, with for example elders or chiefs receiving a disproportionate share, then relative poverty would emerge, but it would not be related to the employment situation, rather to the distribution system. Hence relative poverty is not an appropriate way of defining underemployment in a wholly traditional society.

Of course, this is a highly oversimplified presentation. In traditional societies hours of work vary, incomes vary and productivity varies. The diagram gives a far too homogenised impression. But it is intended to emphasise the point that notions of un- and underemployment do not make much sense in such societies, unless they are used to define the *whole* society as underemployed simply because productivity is lower than in advanced countries. Figure 2.3 presents a stylised picture of what happens when a modern industrial sector is introduced in an underdeveloped economy. This picture is based on a few broad (and fairly crude) generalisations – which are subject to much controversy as to origin and rationale, and qualification as to their total and universal truth – but are very broadly derived from and in accordance with most empirical evidence. The broad generalisations are: *one*, that the modern sector uses technology derived from the advanced countries, with the consequence that the technology has high labour productivity as compared with techniques in use in the traditional sector, and uses relatively investment- and skill-intensive techniques as compared with traditional techniques; *two*, that the sector employs only a small proportion of the workforce of the country – generally under one-fifth; *three*, that incomes generated in the sector – wages and salaries – are substantially higher than incomes elsewhere in the economy. Much of this book is concerned to analyse the realism, causes and consequences of these generalisations. Part has been (implicitly) discussed in Chapter 1. At this stage we simply present the stylised picture in order to illustrate its implications for the definition and analysis of questions of employment, unemployment and underemployment.

In Figure 2.3 work in the modern sector is drawn from the right-hand

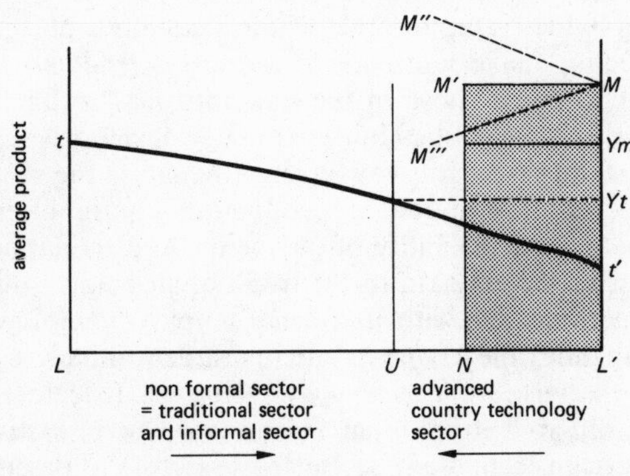


Fig. 2.3

sector, while work in the traditional sector is drawn from the left-hand axis. $LL^1$ is again the total population of working age. $L'N$ are employed in the modern sector with average product shown by $MM'$. (It is assumed that average product is constant as more are employed. Alternative assumptions of decreasing returns ($MM'''$), in accordance with neo-classical assumptions, and increasing returns ($MM''$) in accordance with the, perhaps, more realistic Verdoorn law, are shown by the dotted lines. Which is the correct picture does not affect the argument here though it is of significance for some of the later discussion.) The modern sector wages are $L'Ym$, considerably in excess of traditional incomes. The high rewards of the modern sector create a substantial incentive to switch from the traditional to the modern sector, and one which is accentuated by increasing awareness of possibilities of modernisation, arising with increased education, and the impact of the modern sector throughout society. But the opportunities of employment in the modern sector are strictly limited, to $NL'$ in the diagram. Others, who would have liked to get modern sector employment, have a choice between open unemployment, seeking a job among those that become vacant through turnover, and the increase in job opportunities over time, or remaining in the traditional sector. The traditional sector itself is changed by the industrial sector. On the one hand, opportunities may be limited by competition from the modern sector. Some traditional activities are destroyed, for example, by the modern sector taking away land formerly used by the traditional sector, outlawing many traditional activities and products, e.g. beer-making, and competing away markets for goods and services. On the other hand, the new industrial sector creates new opportunities especially in and around the newly created towns to provide goods and services for the sector, and for those, like the unemployed and those in the urban traditional sector, who are on the fringes of the industrial sector. Those involved in these new activities constitute the 'informal' sector. These new activities which are, on the whole, satellites of the advanced technology sector, make nonsense of the terms *traditional* sector and *modern* sector, because those in the new informal sector are not employed in traditional activities but in recently developed activities. For some purposes, however, it is convenient to analyse the sector with the traditional sector. The mode of production – more often than not self-employment – the flexibility of payments, and the labour-intensive technology used, make it akin to the traditional sector, rather than the so-called modern sector, with its formal employer/employee relationships, high and inflexible payments and investment-intensive technology. We shall use the term *local technology sector* (or L-sector) to describe those in the old-style traditional sector *and* the new-style informal sector; and foreign technology sector (or F-sector) to describe those in the sector using technology developed recently in the advanced countries.

Even these terms are not always accurate descriptions – since the informal sector may often borrow ideas for methods of production from abroad, while occasionally the high-wage sector may use technology developed locally.

The level of open unemployment depends on the extent to which people consider it worth while to spend all their time seeking F-sector jobs: this depends on (*a*) work and income opportunities in the L-sector, both the traditional and informal parts; (*b*) wage rates in the F-sector; (*c*) their estimation of the likelihood of getting an F-sector job as perceived by the potential job seekers, which is a function of education, number of jobs available, the number of other job-seekers, the number of jobs becoming vacant, and employers' criteria for selecting among the unemployed; (*d*) whether it is necessary to be completely without work to secure an F-sector job, or whether job-seeking there can be combined with L-sector work; (*e*) extent (and length of time) for which the unemployed will receive financial support.

The Harris-Todaro model (1970) takes the income differential between sectors, (*a*) and (*b*) above, and the estimated probability of each job-seeker securing a job, (*c*) above, which is assumed to depend on the ratio of jobs to job-seekers, as the critical determinants of the level of open unemployment. The model ignores (*d*) – the possibility that people may seek work without being openly unemployed – and (*e*) as determinants of the level of unemployment.

Let us assume that $NU$ choose to be unemployed. Then $LU$ are working in some capacity in the L-sector; assuming they each receive the average product, their incomes will be $L'Yt$. If incomes are unequally distributed some may receive more than $L'Yt$, some less. When it comes to defining underemployment in such an economy the problem is whether to include all/none or some of those in the L-sector, and if some, where to draw the line. It is clear from Figure 2.3 that there is no natural dividing line in such an economy, any more than there was in the case of the purely traditional economy.

Nurkse provides one apparently non-arbitrary dividing line: include as underemployed those whose marginal product is zero. But, as argued, there are more or less insuperable measurement problems, and in so far as measurement has been made it appears that marginal product is not zero. Yet any other dividing line based on 'low' incomes is arbitrary, depending on how incomes are distributed within the traditional sector, and how low 'low' is defined to be. The same work situation could, with a low-income approach, lead to completely different measures of underemployment according to whether incomes are evenly distributed within the traditional sector or unevenly. In the case of equal distribution, either the whole sector, or none of it, would be defined as underemployed depending on whether the average income was above or below the chosen underemployment poverty line. In the case of unequal distribution the extent of underemployment depends on factors governing the inequality of income distribution. The employment problem is then transformed into an income distribution problem.

### EMPLOYMENT CONCEPTS AND TECHNOLOGY

Emphasis has been placed on the apparently sterile question of definition, because definition and analysis of employment/unemployment in third world countries is of critical importance to assessing the role of technology. We may illustrate this by discussing three types of analysis that have been put forward to explain the employment problems of underdeveloped countries: distorted factor prices,[18] the factor proportions problem[19] and inappropriate technologies.[20]

'*Distorted' factor prices:* these form the central pivot of much neo-classical analysis of unemployment in underdeveloped countries. The argument is that there exists some 'natural' or 'undistorted' set of factor prices – wage rates, interest rates, and prices of investment goods – which would reflect the opportunity cost of the various resources, and which would ensure full employment of all resources. Un- or under-employment is then attributed to various market imperfections distorting the factor prices and thus preventing full employment of resources. For example, trade unions and government wage regulation interfere in the labour market, pushing up wages above their natural rate; subsidies on investment, designed to increase the rate of investment, artificially cheapen it. Similarly, overvalued exchange rates maintain real wages above the natural rate, and the costs of imported investment goods below the natural rate. Figure 2.4 illustrates the situation. There is assumed to be an infinite choice of techniques of varying labour and investment intensity, allowing any investment rate to be associated with any rate of employment. Full employment of resources would occur with price line, $PP'$, given limited investment resources $Oi$, and labour supply $OL$. But distortions of the kind mentioned result in the actual price line of $DD'$. Given limited investment resources, $Oi$, unemployment $uL$ results. The solution to the problem is clear: 'Let the endowment speak', as one author has put it.[21] Factor prices should be changed, or factor use subsidised/taxed to bring them into line with 'undistorted' full employment prices.

A fundamental objection to this approach is the assumption made about the nature of technological choice – both as to the range of techniques assumed, and to the critical determinants of choice. The discussion of technological choice in the last chapter demonstrated the unrealism, in most cases, of the neo-classical assumption of a wide range of available techniques of varying labour and capital intensity to produce any given product, while it also suggested that the relative price of capital and labour was only one among a number of determinants of the choice of technique. The argument will be further developed in later

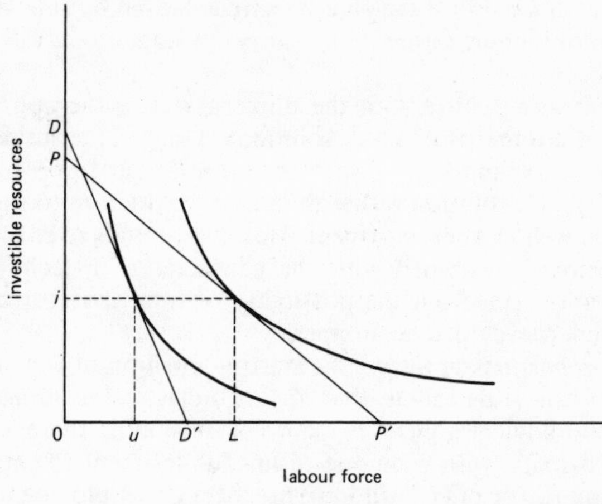


FIG. 2.4

chapters. A second objection is the assumption that wages paid may be freely varied without affecting labour efficiency necessary for its working. The labour force has to be well fed, healthy, punctual and clothed. The sort of wages required to get 'undistorted' factor prices are often well below the sort of level necessary – and are therefore inconsistent with modern technology. Here we are primarily concerned with the analysis of the employment situation on which this type of approach is based. It is assumed – normally implicitly – that only the modern sector counts in terms of employment, and the rest of the population are more or less equivalent to unemployed. Thus the factor price distortions referred to as *the* central problem are the factor prices ruling in the modern sector. Outside the modern sector there is no question of the factor endowment not speaking. Incomes from work are very low, interest rates high. In the traditional sector, activities are as labour-intensive as the technology and human willingness to work permits. But the full employment that is aimed for is the employment of everyone in the F-sector. By ignoring the rest of the economy the approach may be counterproductive. Reductions in wage rates in the F-sector may increase employment there. But it may also reduce employment opportunities in the traditional/informal sectors by a greater extent, since the lower wage rates will enable the F-sector to undercut the L-sector. Any switch in production from the L-sector to the F-sector tends to reduce total employment opportunities because of the greater labour-intensity of activity in the L-sector.[22] Expansion of the F-sector may also increase the rate of open unemployment through increasing the chance of acquiring a modern sector job and hence the incentive to switch jobs. This may be offset by a sufficient reduction in wage-rates, reducing the reward for securing such a job. But if the change were achieved by subsidies on the employment of labour, rather than changes in wage-rates, this offsetting would not occur.

By concentrating entirely on the modern sector, the approach may thus produce counterproductive solutions. Only if the modern sector could really be expanded to absorb the total labour force does it offer the possibility of a solution rather than an accentuation to the problem of poverty as well as unemployment. However, trends over time – to be discussed below – combined with the character of F-technology (see Chapters 1 and 3) rule out the possibility of total F-sector absorption for many underdeveloped economies.

*The factor-proportions problem*: the starting position of this analysis of the employment situation is that the neo-classical assumption of a range of technological policies is incorrect. Assuming there is complete technological fixity, with a unique ratio of investment to employment, then changing factor prices will have no effect on employment. Technological possibilities are not then represented by a smooth curved isoquant as in Figure 2.4, but by rigid proportions shown by a straight ray from the origin as in Figure 2.5. For any given amount of investment, say, *OK*, a unique amount of employment, *Ou*, will be generated, irrespective of factor prices. Thus total employment is determined by the rate of investment. Unemployment results if, given the ratio of investment to employment, the rate of investment falls below that level necessary to ensure full employment. In the diagram with investment *OK*, unemployment *uL* results. Since technology in the modern sector is developed in advanced countries for their factor proportions as shown in Chapter 1, it is likely that it will be too investment-intensive for full employment in underdeveloped countries. (This point is explored more fully in Chapter 3.)

In so far as there is more than one good and/or more than one process, factor proportions may not be rigidly determined, but free to vary according to the ratio in which different goods are produced (or processes used), and the factor proportions associated with the alternative goods/processes. However, the range of production possibilities may still be too narrow to allow full employment. This is shown in Figure 2.5 (*b*) where, if the more labour-intensive techniques are adopted, unemployment *ML* none the less results.

The rigid factor proportions theory suggests two possible solutions to the problem: *one*, expansion of investible resources to encompass as many as possible of the unemployed; *two*, modification of the modern technology in a labour-intensive direction so as to fit with the availability of resources in underdeveloped countries.

The rigid factor-proportions theory is similar to the 'distorted' prices in concentrating entirely on the modern sector, and assuming that anyone outside the sector is virtually unemployed. As with the distorted

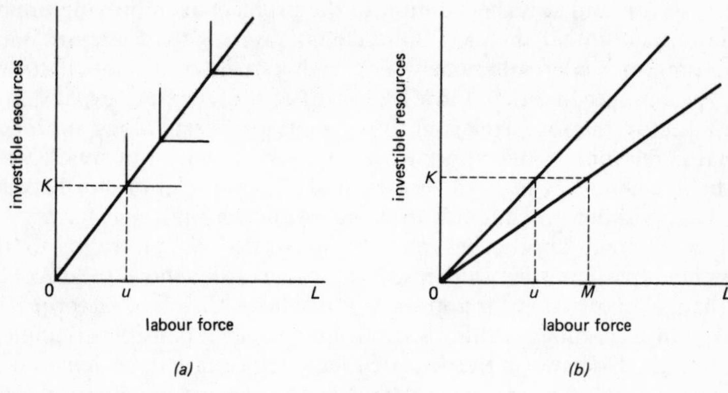


FIG. 2.5

prices theory, the suggested solution may actually increase open unemployment as the F-sector expands and the chance of getting a job increases. Again, the strategy is only feasible if the investment rate can be increased and/or the technology sufficiently adapted to encompass the whole of the labour force.

*Alternative technologies*: whereas the two previous approaches concentrate entirely on the F-sector, aiming to extend it to cover the whole economy, the alternative technology approach is concerned to improve the productivity and work opportunities of the L-sector, as well as modifying technology in the F-sector. The 'distorted' factor price approach and the fixed factor proportions approach both assume that underdevelopment prevails in the L-sector and can only be eliminated by expanding the F-sector. In contrast, the alternative technology approach views activities in the L-sector as just as legitimate as those in

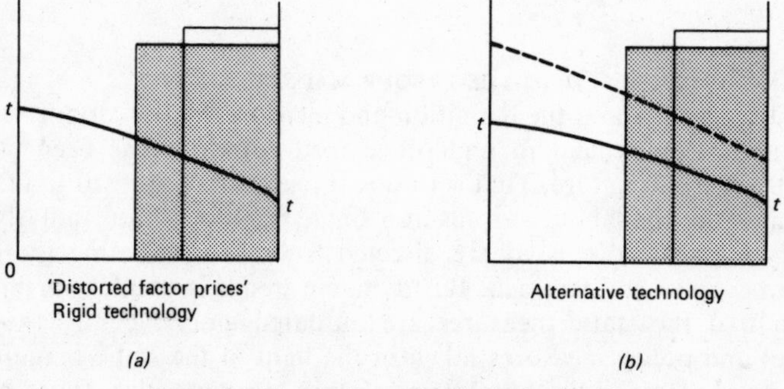


FIG. 2.6 Alternative technological strategies

the F-sector, and sees the solution to the problem as improving opportunities there rather than the full-scale takeover by the F-sector. While measures of underemployment are highly relevant to the first two approaches, because they show the extent of the necessary expansion in the F-sector, they are irrelevant to the alternative technology approach. What is relevant to this approach is an examination of incomes, work and the technology in both sectors, and the interrelationship between the two, without any attempt to define people as *underemployed.*

The different approaches may be illustrated by returning to the previous representation. In Figure 2.6 the shaded area shows the direction of change suggested by the approaches. (*a*) shows 'distorted factor prices' and rigid technology solutions. Both involve an expansion of employment in the F-sector (in the distorted factor price case to be achieved by modifying factor prices, in the rigid factor proportions by modifying F-technology and increasing investment), and a reduction in average productivity in the F-sector (in the distorted factor prices case through inducing a changed selection of techniques in the sector as a result of the modified factor prices, in the rigid factor proportions through new technological developments in the F-sector). In both cases, open unemployment remains more or less the same since the effect of somewhat reduced average rewards is offset by increased F-employment opportunities. Figure (*b*) shows the alternative technology solution. It is two-pronged, involving an increase in employment in the F-sector and a reduction in its labour productivity, like the other approaches, to be achieved by modification of technology in that sector; but it also involves raising productivity in the L-sector. The gap between the sectors is reduced and the consequence is that open unemployment may be reduced since there is a reduced incentive to switch between the sectors, with the greater productivity of the L-sector.

Characteristics of technology in both sectors, and the possibility of moving in the directions suggested, are discussed in much greater detail later. Here we are primarily concerned to show the relationship between definition and analysis of the employment problem, and the analysis of technology.

### A DYNAMIC APPROACH TO THE LABOUR MARKET

The discussion about the definition and measurement of labour surplus and underemployment in underdeveloped countries has been almost entirely static in nature. That is to say, it has been concerned to identify and measure the labour surplus at a single point of time. Similarly, the policy measures discussed are directed towards eliminating the static labour surplus so identified. But dynamic trends in the labour market mean that the static measures are outdated almost before they are made; and policy measures taken in the light of the static magnitudes may be rendered futile by dynamic trends. In particular, those policy measures aimed at expanding the large-scale modern technology sector (the F-sector) until it encompasses the whole economy, are, for many countries, virtually ruled out because of dynamic trends. This section briefly examines these dynamic variables and their effect on the distribution of labour between the F-sector and the L-sector.

The F-sector may be assumed to have no labour supply problem. Because it pays wages considerably above earning opportunities elsewhere, it may employ as much labour as it wishes. (There may be supply limitations of skilled labour, and even supply limitations on the 'right' kind of labour with factory attitudes towards work, but generally these limits do not arise from limited willingness to work in the sector.[23]) The L-sector is therefore in one sense the residual sector providing for all those who do not get employment in the F-sector, and do not spend their whole time seeking employment in the F-sector (the unemployed). Ignoring the unemployed, the size of the L-sector then depends on how many people are left after the F-sector has made its pick, and this depends on the size of the F-sector and the size of the total population of working age. The change in the size of the two sectors over time depends on (*a*) the rate of increase in employment in the F-sector; (*b*) the rate of increase in the population of working age; and (*c*) the initial relative size of the two sectors.

The problem of labour surplus is often attributed to the combined effects of (*a*) and (*b*) – i.e. the relatively slow rate of growth of employment opportunities in the modern sector, and the rapid growth of population. In fact the third variable – the relative numbers in each sector – is of crucial importance in determining the change in the distribution of labour between L and F sectors. A simple example illustrates the point. Take two countries, A and B, and assume that in both countries employment opportunities in the F-sector grow at 5 per cent p.a. and the labour force grows at 3 per cent p.a. Suppose that initially A the F-sector employs 80 per cent of the total labour force, while in B the F-sector accounts for 20 per cent. The initial pattern of employment, with a labour force of 100 in both countries, will be as follows:

| | *Employment in* | |
|---|---|---|
| | *F-sector* | *L-sector* |
| A | 80 | 20 |
| B | 20 | 80 |

After one year the pattern of employment will be

| | *Absolute numbers* | | *% of labour force* | |
|---|---|---|---|---|
| | *F* | *L* | *F* | *L* |
| A | 84 | 19 | 81·5 | 18·5 |
| B | 21 | 82 | 20·2 | 79·8 |

The same percentage rate of growth of employment and labour force has led to a *decrease* in the number of people in the L-sector in A, and an *increase* in B, simply because of the initial relative size of the two sectors. (It is assumed that anyone who is not employed in the F-sector is employed in the L-sector, which therefore should be interpreted to include the unemployed.)

The determinants of the change in numbers in the L-sector can be described more formally:

Let $P$ = total labour force;
$p$ = annual percentage increase in labour force;
$Ef$ = total employment in the F-sector;
$f$ = annual percentage increase in employment in the F-sector;
$Qf$ = the proportion of the total labour force employed in the F-sector, which is, of course, equal to $Ef/P$.

If there is to be no increase in the absolute numbers employed in the L-sector the following conditions must be met:

$$\Delta Ef \geqslant \Delta P, \text{ or since}$$
$$\Delta Ef = f \times Ef$$
$$\Delta P = p \times P,$$
$$f \times Ef \geqslant p \times P,$$
$$\frac{Ef}{P} \geqslant \frac{p}{f} \quad \text{or} \quad Qf \geqslant \frac{p}{f} \qquad (1)^{24}$$

That is if the numbers in the L-sector are not to grow the proportion employed initially in the F-sector must be at least as great as the percentage increase in the labour force divided by the percentage increase in employment opportunities in the F-sector. For a 5 per cent annual increase in employment in the F-sector and a 3 per cent annual increase in the labour force the proportion employed initially in the modern sector must be at least 60 per cent if numbers in the L-sector are not to grow. Inequality (1) can be interpreted in a different way to show the minimum rate of growth of F-sector employment that would be required, given the rate of growth of the labour force and the distribution between the two sectors, if the L-sector is not to grow. Taking the example of India with a $Qf$ of 6·8 per cent and a rate of growth of labour force of around 2 per cent p.a., the rate of growth of employment opportunities must be at least 28 per cent p.a. if absolute numbers outside the modern sector are not to increase. In the past the rate of growth of F-sector employment has been around 4 per cent p.a.[25]

Over time the proportion of people employed in the F-sector will change so long as $p \neq f$. In year 2,
$(Qf)_2$ will be

$$\frac{Ef + \Delta Ef}{p + \Delta P}$$

or

$$\frac{Ef(1+f)}{P(1+p)}$$

or

$$(Qf)_1\frac{(1+f)}{(1+p)}$$

In year $t+n$,

$$(Qf)_{t+n} = (Qf)_1\left[\frac{1+f}{1+p}\right]^n$$

Numbers outside the F-sector (i.e. those unemployed or in the L-sector) will cease to grow when

$$(Qf)_{t+n} = \frac{p}{f}$$

The number of years, $n$, this will take to occur is given by the equation,

$$(Qf)_1\left[\frac{1+f}{1+p}\right]^n = \frac{p}{f}$$

If we know the initial proportion in the F-sector and the rates of growth of employment and the total labour force we can calculate the number of years before the L-sector can be expected to stop growing. (This will only happen if $f > p$; if $f < p$, $Qf$ will decline over time and the numbers in the L-sector will grow at an increasing rate.) To go back to the example of India with rates of growth of labour force and employment shown in Table 2.3 it will take 100 years before numbers outside the F-sector stop increasing.

The problem about making any empirical estimates of these values is that data is not collected for the F- and L-sectors – indeed it is difficult to know precisely where to draw the dividing line between the sectors. Moreover, variations in the values of the variables $Qf$, $f$ and $p$ can make a substantial difference to the results. Probably estimates which come closest to measuring employment in the foreign technology sector are those of Doctor and Gallis for some Asian countries; they exclude employment in small-scale establishments, domestic servants and all agricultural employment except for plantation employment in their estimates for India, shown in Table 2.3. However, their estimates are now rather out of date. The Doctor and Gallis estimates for Africa include all wage employment. In the table they have been used excluding agricultural employment. It is probable therefore that the estimates in Table 2.3 (other than those for India) exaggerate the size of the F-sector, and therefore of $Qf$. Whether and how this affects the final results, shown on the right-hand side of the table, depends on whether the rate of growth of employment in the F-sector would be higher or lower using a corrected definition. The Doctor and Gallis estimates for India, while they show a smaller F-sector, also show a higher growth in

TABLE 2.3 Some empirical estimates

| Country | (1) $Qf$ (year) | (2) $f$ (years) | (3) $p$ I.L.O. projection 1965–70 | (4) [$p$] actual population growth 1963–72 | (5) $\frac{p}{f}$ | (6) $\bar{f}$ | (7) $n$ |
|---|---|---|---|---|---|---|---|
| Egypt | 29·7[a] (1960) | 1·6 (68–71) | 2·3 | [2·5] | $p>f$ | 7·7 | moving away |
| Gabon | 14·1[a] (1963) | 8·0 (64–69) | 0·4 | [1·1]* | 5 | 2·8 | already achieved |
| Ghana | 14·3[a] (1960) | 0·5 (64–70) | 2·8 | [2·9] | $p>f$ | 19·6 | moving away |
| India | 6·8[b] (1961) | 3·9[b] (51–61) | 1·9 | [2·2] | 49 | 27·9 | 100 |
| Kenya | 16·3[c] (1970) | 3·6 (66–71) | 2·8 | [3·3]* | 78 | 17·2 | 201 |
| Malawi | 5·8[c] (1970) | 8·6 (68–72) | 2·2 | [2·8]* | 26 | 37·9 | 25 |
| Mauritius | 17·4[c] (1970) | 5·4 (64–72) | 2·8 | [2·0] | 52 | 16·1 | 44 |
| Philippines | 23·5[d] (1961) | 4·1[d] (56–61) | 2·8 | [3·0] | 68 | 11·9 | 83 |
| Sierra Leone | 5·7[a] (1963) | 1·2 (64–71) | 1·7 | [1·5] | $p>f$ | 29·8 | moving away |
| Turkey | 24·5[a] (1970) | 9·2 (64–72) | 2·1 | [2·5] | 23 | 8·6 | already achieved |
| Uganda | 6·4[c] (1970) | 6·7 (64–71) | 2·2 | [2·8] | 33 | 34·4 | 38 |
| Tanzania | 5·0[c] (1969) | 5·5 (64–70) | 2·3 | [2·6] | 42 | 46·0 | 70 |
| Zambia | 27·6[a] (1961) | 7·2 (64–71) | 2·6 | [2·6] | 36 | 11·5 | 10 |

NOTES AND SOURCES

(1) $Qf$ is supposed to represent the proportion of the labour force employed in the foreign technology sector. Because statistics are not available which precisely measure this, as a second best the figures are:

[a] total wage employment excluding agriculture as a proportion of the labour force (Doctor and Gallis (1966)).

[b] Doctor and Gallis (1964) estimates for 'modern sector' employment as a proportion of total employment. This excludes employment in smaller-scale establishments, domestic servants and all agricultural employment except in plantations.

[c] Wage employment excluding agriculture as a proportion of the labour force (United Nations, *Survey of Economic Conditions in Africa, 1971*).

[d] Doctor and Gallis (1964) estimates for 'modern sector' employment, as in (b) above, but not excluding smaller scale establishments.

(2) $f$ : annual growth in employment excluding agriculture (*United Nations Statistical Year Book, 1973*).

(3) $p$ : International Labour Office projections of annual growth in labour force, 1965–70. (I.L.O., *Labour Force Projections 1965–1985*).

(4) $[p]$ : actual annual growth in total population 1963–72 (*United Nations Statistical Year Book, 1973*);* projected for ages 15–64, 1970–80, (*Survey of Economic Conditions in Africa*), 1971.

(5), (6) and (7) are calculations based on the figures in (1), (2) and (3) in the table.

(5) shows required $Qf$ if expansion of F-sector is to provide jobs for all additions to the labour force on current trends in labour force increase and employment increase.

(6) $\hat{f}$ shows required rate of increase of F-sector if it is immediately to absorb *all* additions to the labour force.

(7) $n$ shows number of years, on current trends, before the F-sector will absorb all additions to the labour force.

employment than on an alternative base, and hence to some extent there are offsetting changes in making the final calculations. Table 2.3 uses the International Labour Office projections of the labour force for 1965–70. Column 4 shows past rates of increase of population which are likely, broadly, to be reflected in labour force increases in subsequent periods. For the most part these are higher than the I.L.O. projections, suggesting that the calculations on the right hand-side of the table may be underestimates.

Precise estimates are impossible to obtain. They are in any case unimportant. The significant point that is illustrated by the calculations is that definition and measurement of *static* employment situations may be dwarfed by dynamic trends. Table 2.3 shows that for most countries the F-sector will not absorb additions to the labour force for many years. Until this point is reached it cannot begin to absorb any underemployed from those already in the L-sector. Any proposed 'solution' to the employment problems in many underdeveloped countries must take this dynamic situation conclusion, derived from the dynamic situation, as the starting point.

NOTES

1. Among the vast and growing literature on the subject that bears witness to this see for example Jolly *et al.* (1973), the I.L.O. reports on employment in particular countries, Turnham's report for the O.E.C.D., Edward's report for the Ford Foundation (1974), H. Bruton's survey article (1973) and Morawetz's (1974) survey article.
2. This problem is discussed more fully in Streeten (1970), Myrdal (1968) and Stewart (1970).
3. Walter Elkan (1960) provides a fascinating case study of transitional modes, where wage labour is a temporary supplement to subsistence activities.
4. See Boserup (1965 and 1970), who illustrates how the employment relationship in agriculture alters according to environmental differences, and similarly how that of women differs between societies.
5. See e.g. Hart (1971), I.L.O. (Kenya) (1972).
6. See Hutton (1966) and Pfefferman (1968).
7. Turnham (1971) supports this conclusion empirically.
8. Joan Robinson (1937, p. 84) originally coined the term for developed countries in cyclical depression: 'it is natural to describe the adoption of inferior occupations by dismissed workers as *disguised unemployment*'.
9. Sen (1975), Appendix A, uses a similar diagram.
10. See Rado's explosive model of education (1973), and Blaug *et al.* (1969) on graduate unemployment in India.
11. For example, Turnham (1971) shows that 19·5 per cent of the economically active population had incomes below the statutory minimum wage of the time, in the Lima-Callao metropolitan area, and were working over 35 hours a week (p. 70). The I.L.O. Report on Colombia found 12 per cent of the labour force were as poor as those who were totally without work, despite working over 32 hours a week. See Krishnamurty (1975) for similar evidence on India.
12. Nurkse (1953), p. 32.
13. For evidence of positive marginal product see Paglin (1965), Bennett's comment and Paglin's reply in the *A.E.R.*, research reported on in Kao, Anschel and

Eicher (1964), and other work referred to in Turnham (1971), p. 68, n. 1. Desai and Mazumdar (1970) find zero and negative marginal product per work-hour among some family farms.

14. Nurkse (1953), p. 33.
15. *Measurement of Underemployment, Concepts and Methods*, I.L.O. Report IV, Annex III, p. 90.
16. Weeks (1971).
17. Here, and in what follows, we use the concept 'population of working-age' rather than 'labour force'. The labour force normally includes only those who work outside the home. In all economies it is irrational (and insulting) to exclude work in the home from work statistics. In underdeveloped countries where family activities include many, often all, income-creating activities the distinction is particularly misleading.
18. The familiar neo-classical diagnosis – see, for example, Harris and Todaro (1969), and Ranis (1971).
19. Described by Eckaus (1955).
20. See for example, Schumacher (1973), Marsden (1971) and Dickson (1974).
21. Ranis (1969).
22. The conditions in which this may occur are discussed in Stewart and Weeks (1975).
23. This has not always been the case. When industrialisation first occurred the new factories had considerable difficulty in attracting labour, and had to use compulsion of various kinds: for a graphic description of early labour supply problems in Kenya, see E. Huxley. See also the Carpenter Report on African Wages (1954) and Weeks on Nigeria (1971).
24. Since developing this approach I have discovered similar approaches to the question of agricultural labour force in F. Dovring (1959) and in Johnston (1966).
25. See notes to Table 2.3 for sources.

# 3 Inappropriate Technology

Earlier (Chapter 1) we discussed the determinants of technological choice. This chapter is concerned to show how the historical development of technology described there has led to the use of inappropriate technology in the third world, and the implications of this for patterns of development. The next chapter considers the characteristics of what is described as 'appropriate' technology.

Broadly (and with a few notable exceptions), technological developments over the past 150 years have been concentrated in what we describe as the advanced countries. The vast bulk of the world's capital goods manufacture is also concentrated there. This means that, in addition to traditional techniques, third world countries have little choice but to use techniques developed in advanced countries, and for the most part produced there too. Consequently, the available technological choice is largely circumscribed by technological developments in the advanced countries. While sales of second-hand machinery and capital goods capacity in third world countries enable the third world to make some use of techniques now obsolete in advanced countries, technological choice is mainly confined to techniques *currently* produced in the advanced countries – i.e. to the more recently developed techniques. Reinforcing the bias against older techniques is the fact that many of them have become obsolete directly or indirectly, for the third world as well as the advanced countries. As shown in Chapter 1, scientific and technical progress may make older techniques and products technically inefficient, using more of all types of resources than later techniques. This is described as *direct* obsolescence. *Indirect* obsolescence occurs when other changes in the system make a particular technique obsolete – e.g. input requirements change so that a given product becomes obsolete.

Product changes play an extremely important role in influencing technological choice towards the latest techniques. Product changes may lead to more efficient products, with a consequent waste of resources if older products are used; moveover, the many technical linkages – of production and consumption – mean that in large part the technology package has to be accepted (or rejected) as a whole – part selection is impossible. Thus a change in technique in one part of the system requires changes in product design (and hence technique) of linked processes/products.[1] In many economies the nature of demand is such that firms have to keep abreast with recent product developments to sustain their market share. This is most obviously true in export markets, but it is also true in domestic markets where oligopolistic competition between firms largely takes the form of product competition, supported by advertising and other forms of promotion, rather than price competition. In such economies, firms which fail to keep their products and processes up-to-date tend to lose out to those that do. Langdon (for the soap industry in Kenya) and Sercovich (for a number of industries in Argentina) have provided empirical case studies of this phenomenon.[2]

Moreover, within the limited technological choice available, there is a strong tendency, in the modern or F-sector, for later and more capital-intensive techniques to be selected. This is partly because the later techniques are, as argued, more efficient. But it is also because the selection mechanisms – i.e. the determinants of choice – tend to be biased towards the selection of such techniques. The nature and consequence of such bias will be discussed more below. The important point here is that the selection mechanisms accentuate the trend towards the use of advanced-country technology, which is itself technologically dictated – i.e. is the consequence of world technological developments. By reinforcing this trend the selection mechanisms appear to *justify* the use of advanced-country technology, making it appear privately and often socially profitable. This has two consequences: on the one hand, it means that there is little incentive to search for an alternative technology, or a modification to the advanced-country technology – since the extant advanced technology appears justified; on the other, it concentrates the attention of those dissatisfied with the consequence of the technological choice on changing the selection mechanisms, rather than searching for alternative technologies.[3] Concentration on the selection mechanisms might make sense if the selection mechanisms were independent of the technology in use. But the selection mechanisms are not independent of the technological choice but themselves largely flow from it, being part of the technology package. Thus the fact that the selection mechanisms appear to justify the use of advanced-country technology may simply reflect the self-justifying character of the technology package, with the use of advanced-country technology leading to a consistent set of selection mechanisms. This will be discussed more below.

In Chapter 1 technology was described as a matrix consisting of a set of techniques, each of which was associated with a vector of characteristics. These characteristics include the nature and specification of the product, the input use, the scale of production, associated managerial techniques, etc. The characteristics are designed to fit in with the economic, institutional and technical circumstances of the economy for which they were designed. Techniques developed in advanced countries have characteristics largely conditioned by the economic environment in the advanced countries. The inappropriate nature of advanced-country techniques for underdeveloped countries arises from differences in the economic and institutional environment between advanced and third world countries. As a result of these differences, the techniques from the advanced countries have inappropriate characteristics for underdeveloped countries: when they are transferred to third world countries, their economies are distorted in the effort to reproduce the sort of conditions for which the techniques were created. In so far as the effort to reproduce these conditions is costly and only partially successful, the transfer of the techniques leads to inefficiency, as well as distortions.

There are obvious climatic differences between advanced countries and third world countries – of temperature, humidity, seasons – which lead to differences in natural vegetation, in conditions of cultivation and of production. These differences affect agriculture most, and few would attempt to transfer agricultural techniques unaltered from one type of country to another. Climatic differences also affect processing, manufacturing and construction technology. Most obviously, they may result in different availability and costs of raw materials. In addition, the processes may be designed for and only work successfully under particular conditions of heat and humidity, and the appropriate atmosphere may have to be reproduced – by temperature control, etc. – and/or inefficient production accepted, when the process is transferred to a different climate. In all sorts of respects, technology designed in advanced countries does reflect their climatic conditions – with the consequence of increased cost/decreased efficiency when it is transferred. For example, paper manufacture has been designed using conifers as raw material, and only recently has research been concentrated on using the materials widely available in underdeveloped countries. Modern techniques of energy production are focused on conditions in advanced countries, resulting in, e.g., concentration on nuclear energy rather than on sources of energy like sun and wind which are abundantly available in third world countries. These physical differences alone mean that the advanced-country technology is often inappropriate. While the physical differences are most obvious, most is being done about modifying technology to allow for them. But here we are more concerned with economic and institutional differences that make advanced-country technology inappropriate.

Chapter 1 classified the environment which formed techniques into three aspects: institutional factors, or organisation of production; economic factors, particularly income levels; and technical factors. We shall consider the relationship between conditions in advanced countries and poor countries in the same way.

*Organisation of production*: two or three hundred years ago, organisation in the now advanced countries was probably not all that dissimilar to conditions in large parts of the underdeveloped countries today: traditional employment ties of a feudal or semi-feudal nature operated side-by-side with self-employed units; as the latter expanded they gradually began to take on additional employees from among those

TABLE 3.1 Trends in industrial concentration

U.K.

| *% of total employees employed in plants of 1000+ employees* | *1935* | *1958* | *1972* |
|---|---|---|---|
| Food, drink and tobacco | 24 | 31 | 31 |
| Chemicals | 22 | 42 | 40 |
| Metal industries | 28 (750+) | 42 (750+) | 33 |
| Engineering and shipbuilding | 40 | 48 | 40 |
| Motors | 63 (750+) | 77 (750+) | 66 |
| Aircraft | 77 | 80 | 80 |
| Textiles | 19 (750+) | 15 (750+) | 16 |
| Clothing | 19 (500+) | 18 (500+) | } 5 |
| Footwear | 15 | 7 | } |
| Bricks etc. | 10 | 21 | 21 |
| Timber etc. | 10 (400+) | 12 (400+) | 9 (400+) |
| Rubber | 60 | 53 | 49 |
| Paper manufacture | 11 | 21 | 16 |
| Printing | 19 | 23 | 22 |
| All manufacturing | 22 | 35 | 33 |

SOURCE Armstrong and Silberston (1965), *Census of Production*, 1963, 1972.

U.S.

| *% share of 500 largest corporations in* | *1960* | *1973* |
|---|---|---|
| Net sales of manufacturing corporations | 59 | 65 |
| Net profits after tax of manufacturing corporations | 80 | 79 |
| Employment of manufacturing corporations | 55* | 76 |

* % of all manufacturing employment

SOURCE *Fortune*, 1961, 1974; *Statistical Abstract of the U.S.*

TABLE 3.2 Variation in production cost in relation to different scales of output in selected industries

| *Product, capacity and cost* | *Unit* | *Variation in capacity and production cost* | | | |
|---|---|---|---|---|---|
| Steel Capacity | Thousands of tons per year | 50 | 250 | 500 | 1000 |
| Cost per ton | 1948 U.S. dollars | 209·4 | 158·8 | 137·5 | 127·2 |
| Cement | | | | | |
| Capacity | Thousands of tons per year | 100 | 450 | 900 | 1800 |
| Cost per ton | 1959 U.S. dollars | 26·0 | 19·8 | 16·4 | 13·9 |
| Ammonium nitrate | | | | | |
| Capacity | Short tons per day | 50 | 100 | 150 | 300 |
| Cost per ton | 1957 U.S. dollars | 190·4 | 145·1 | 125·6 | 101·5 |
| Beer bottles | | | | | |
| Capacity | No. of moulding machines | 1 | 2 | 6 | 12 |
| Cost per gross | 1957 U.S. dollars | 8·51 | 7·25 | 6·13 | 5·69 |
| Glass containers | | | | | |
| Capacity | No. of moulding machines | 1 | 2 | 6 | 12 |
| Cost per gross | 1957 U.S. dollars | 8·66 | 7·77 | 6·78 | 6·33 |
| Radial ball-bearings | | | | | |
| Capacity | Production index (1961 = 1) | 1 | 2 | 3 | |
| Cost per thou. | 1961 yen | 79,800 | 67,100 | 63,100 | |
| Tar | | | | | |
| Capacity | Tons per day | 100 | 200 | 300 | 400 |
| Cost per ton | Thousands of 1961 yen | 10·5 | 9·6 | 9·2 | 8·9 |
| Benzol | | | | | |
| Capacity | Tons per day | 50 | 100 | 200 | 300 |
| Cost per ton | Thousands of 1961 yen | 29·2 | 27·1 | 25·9 | 25·4 |
| Aluminium plate | | | | | |
| Capacity | Tons per year | 200 | 1200 | 3000 | 5000 |
| Cost per ton | Thousands of 1961 yen | 276·8 | 272·2 | 269·1 | 263·5 |

SOURCE *Industrialisation and Productivity*, No. 8, op. cit, Table 1.

released from the land and their feudal obligation by the agricultural revolution. These expanding units introduced wage-labour as an important mode of production, and formed the basis of modern employer–employee relationships and modern firms. Historians have now established that there were many forms of employment in the pre-industrial revolution era in Europe. Equally, there are major differences between pre-industrial revolution Europe and underdeveloped areas today.[4] But the differences are probably less than those between today's industrial structure in the advanced country and the pre-advanced country technology forms in underdeveloped countries. In the advanced countries, firms based on employer–employee relationships largely replaced both feudal and self-employment modes. The firms became increasingly large-scale and increasingly specialised.

Table 3.1 illustrates the increasing size of firms and plant in recent years in the U.S. and U.K. In Britain in 1968 the largest 50 firms accounted for nearly one-third of total output and 30 per cent of total employment. Plants employing more than 1000 people accounted for one-third of the total number of employees in manufacturing industry.

The increase in scale and specialisation of function was both cause and product of technical changes. Technical changes make increasing scale profitable. And as the increasing scale took effect, further technical changes were designed for the new forms of organisation, so that they could only be operated by large and specialised units. Large firms account[5] for a more than proportionate share of research and development, devoting the expenditure to technology suited to their form of organisation. Thus the relative advantages of such forms increase, until alternative forms become unthinkable – since the only sources of techniques suitable to small-scale units are from the early nineteenth century, to be found only in science museums, requiring materials that are no longer available, using skills long since forgotten and producing goods that people no longer consume.

The existence of economies of scale – up to a pretty large scale – in many modern industrial processes is well established.[6] Some examples are contained in Table 3.2.

Pratten estimated the minimum efficient scale[7] of different types of production as a percentage of the U.K. market in 1969. Table 3.3 summarises his results.

With much lower per capita income, the size of the market in most underdeveloped countries is only a fraction of that of the U.K. market, as Table 3.4 shows.

Thus even those industries which are efficient at a relatively small scale (e.g. 1 per cent of the U.K. market) are large in relation to the markets of most underdeveloped countries. The consequence is that in many lines one plant is sufficient, often more than sufficient to cater for the entire market. Production therefore tends to be monopolistic and

TABLE 3.3 Estimated minimum efficient scale as a % of the U.K. market in 1969

| (manufacturing) | *No. of industries* |
|---|---|
| 1% of the U.K. market and under | 8 |
| 1–5% | 5 |
| 5–20% | 6 |
| 20–50% | 7 |
| 50% and over | 12 |

SOURCE Pratten (1971).

TABLE 3.4 Manufacturing as a proportion of the U.K. market, 1971

| *Country* | *Value-added in manufacturing as % U.K.* | *Employment in manufacturing as % U.K.* |
|---|---|---|
| Kenya | 0·3 | 1·0 |
| Malawi | 0·1 | 0·3 |
| Nigeria | 1·3 | 1·9 |
| Colombia | 2·8 | 4·2 |
| Ecuador | 0·4 | 0·6 |
| Peru | 2·6 | 1·7 |
| Iraq | 0·4 | 1·3 |
| Jordan | 0·1 | 0·1 |
| Philippines | 2·2 | 5·4 |
| Singapore | 1·0 | 1·9 |
| Tunisia | 0·3 | 0·7 |
| Uganda | 0·1 | 0·6 |
| Brazil | 16·3 | 26·1 |
| India | 6·5 | 57·2 |

SOURCE *U.N. Statistical Year Book*, 1974.

centralised. The large scale, in relation to the size of the market, is one of the causes of the excess capacity to be found in many industries in many poor countries. For example, a survey in W. Pakistan in 1965–6 suggested that the average industrial firm operated its equipment only *one-third* of the time.[8] Similar evidence of low-capacity utilisation has been produced for other countries, despite the apparent shortage of investment resources.[9]

The huge gulf between the organisational form to which advanced-country technology is designed and suited – where a plant employing over 1000 workers is typical – and the forms of organisation indigenous to most third world economies – where family enterprises employing say up to 20 people are notably large and successful[10] – has meant that the use of advanced-country technology also generally requires the use of advanced-country organisations. These organisations are directly imported in the case of foreign investment; in other areas they are imported indirectly via management contracts and/or training managers overseas. Advanced-country technology thus leads to advanced-country techniques of management. It rules out the possibility of local development of local entrepreneurial talent – the jump between current experience and what is needed is far too great. The result appears to be a chronic shortage of entrepreneurial talent: some attribute lack of development to this so-called shortage.[11] But the underlying cause of the shortage is the result of the import of a technology which requires managerial techniques completely out of line with local possibilities. Managerial dependence follows from technological dependence. And of course technological dependence is reinforced by managerial dependence, since the imported managers/managerial techniques are trained/designed to manage the imported advanced-country techniques and therefore when they make a technical choice they do so in directions in which their specialised advantage lies – in the further import of advanced-country technology.[12]

Some believe economies of scale to be invincible technological *truths* – facts of life, technologically speaking. And within existing technology they are: to produce on a smaller scale may involve losses in output which are often substantial, but the economies of scale are themselves formed by the history of technological development, in which particular organisational forms have favoured the development of large-scale techniques. In the competitive struggle, firms have to grow in order to survive. Market imperfections have added a premium on to size, additional to and independent of production technology and production economies of scale: thus advertising has introduced a new source of overhead and scale economy, while market power tends to increase with size, adding to the advantages of size. The growth in the size of the firm can be explained in terms of the competitive struggle between firms combined with market imperfections. It also undoubtedly has a technological counterpart, in economies of scale and specialisation. But these are themselves in part the product of the economic system, in which large firms dominate, which gave birth to them. In an alternative system in which small units dominated the process of technological development, techniques suited to such units would be developed. Today, it is impossible to decide how much of the undoubted economies of scale are due to the organisation of the economy and therefore of technological development, and how much to genuine physical facts which must be present in any organisational system. What is known is that very little of modern scientific and technical effort has been devoted to producing techniques suitable for small-scale units, and that where minification has been attempted[13] its success has surprised engineers and economists, who have assumed that what is, must be.

*Income levels*: the most obvious and stark difference between advanced countries and underdeveloped countries is in income per capita. This indeed, despite difficulties of estimation, comes nearest to the defining difference between the two sets of countries. Table 3.5 illustrates this.

Income levels, as argued in Chapter 1, are of great significance in forming technological development, from the point of view both of demand and of supply. Both the balance (the proportion of income spent on different categories of good) and the nature of demand are dependent on the income levels of consumers. On the supply side, income levels per worker represent (roughly) the opportunity cost of labour, so that the required labour use and labour productivity of the technology are dependent on this cost. In addition, resource availability – in terms of savings per capita, and educational resources per capita – are closely related to income per capita. When techniques designed for one income level are transferred to countries with quite different income levels, it has distorting effects on the economy as a whole. We shall examine these different aspects in more detail.

TABLE 3.5 Gross National Product per capita ($) in selected countries, 1973

| Country | GNP per capita ($) |
|---|---|
| Australia | 4925 |
| Canada | 5377 |
| France | 4892 |
| Germany | 5624 |
| U.K. | 3172 |
| U.S. | 6155 |
| Tunisia | 430 |
| Botswana | 220 |
| Chad | 70 |
| Nigeria | 170 |
| Zambia | 390 |
| Mexico | 810 |
| Argentina | 1420 |
| Brazil | 600 |
| Burma | 90 |
| India | 120 |
| Sri Lanka | 110 |
| Korea Rep. | 370 |
| Philippines | 250 |

SOURCE *Development Cooperation, 1975 Review* (O.E.C.D., 1975), Statistical Annex, Tables 46 and 48.

Let us start on the supply side. Technological developments occur against the background of the resource availability of the economy for which they are intended. If a new technique would require resources which are not available, or only available on a minute scale, then it is unlikely to be developed. Availability or non-availability is not normally a clear-cut matter: resources may be scarce but available at high cost, or they may require training/technical research and development before they are abundantly available. In such cases, the cost of acquiring/developing the required resources has to be taken into account in decisions about whether to develop the new technique. Thus the resource use of a new technique takes into account existing supplies of various resources and their costs, and the cost of developing new supplies. The relative cost of different types of resources is also significant in determining what type of technique, with what type of resource use, is most worth developing.

The technology in use in the economy in question, combined with economic factors determining relative resource costs, thus provides the parameters determining the type of techniques likely to be developed, and once developed successfully introduced.

We have been talking about 'resources' in rather vague terms. What is in fact in question is the inputs used by the techniques – viz. the quantity and quality of labour requirements, the quantity of machinery and the material inputs. A new technique may use new machinery or require new skills of labour or new materials, in which case it will be part of the development of the techniques to provide these, but existing supplies – quantitatively, qualitatively, and their prices – will heavily influence which techniques are likely to be viable, which not. The availability and price of material inputs is part of the technology in use, which will be discussed more below under 'technical factors'. Here we are concerned with labour and machinery. The price, the quality and the availability of both these inputs are closely related to the income level of the economy.

Consider labour. Any technique must pay the ruling wage-rate. This means that the real wages in the rest of the economy help determine which types of technique are viable. As real wages rise, new techniques have to show increasing levels of labour productivity to succeed. The determinant of the wage share in income is a controversial question which has exercised economists since economics began, and will not be discussed here. But irrespective of the determinants of the share, it is clear that absolute levels of real wages vary over the longer period with absolute levels of labour productivity. Hence the higher labour productivity in the economy as a whole, the higher the requirement for labour productivity of new techniques. In general, then, societies with high real incomes will tend to develop techniques with correspondingly high labour productivity. A large part of the process of obsolescence consists in new techniques with higher than average labour productivity, having the effect of raising average wage rates so that older techniques, with lower labour productivity, become unprofitable and are scrapped.

The developed countries have experienced rising labour productivity and rising real wages over the past 200 years, as shown for Germany, the U.S. and the U.K. in Table 3.6. The techniques developed and introduced in the developed countries both reflect and contribute to this trend. Any case study of the development of particular technology in a particular industry shows the same trend: for example, the case of textile machinery discussed below.

TABLE 3.6 Trends in real wages and industrial labour productivity

| | *Germany* | | *U.K.* | | *U.S.* | |
|---|---|---|---|---|---|---|
| | *I.P.* | *R.W.* | *I.P.* | *R.W.* | *I.P.* | *R.W.* |
| 1860 | 50 | 67 | n.a. | 53 | n.a. | 68 |
| 1880 | 78 | 77 | 69 | 73 | n.a. | 72 |
| 1900 | 112 | 108 | 104 | 104 | 105 | 110 |
| 1920 | n.a. | n.a. | n.a. | 104 | 129 | 151 |
| 1938 | 202 | 155 | 141 | 133 | 183 | 203 |
| 1960 | 248* | 282 | 171 | 219 | 368 | 381 |

I.P. = industrial labour productivity – in Germany includes commerce and finance: 1890–9 = 100.
R.W. = index of industrial real wages: 1890–9 = 100.
*1959

SOURCE Phelps-Brown (1973).

The inverse of (relatively) high labour productivity is (relatively) low labour requirements per unit of output. In part reduced labour requirement per unit of output, like reductions in other input requirements, is a natural consequence of scientific and technical progress leading to an overall increase in productivity. In part also it is the result of the substitution of other resources for direct labour power, particularly the increasing use of more – and of more powerful, elaborate and sophisticated – machinery. This has been made possible by the increasing level of savings that has accompanied rising incomes.

The exact relationship between savings and income per capita is not known, but it has long been established that there is a positive relationship between the two variables, so that, in general, the higher incomes per head, the higher savings per head. Historically, as incomes per head rose, savings per head rose so that the resources available for investment per head rose also. Later techniques developed in the advanced countries therefore use increasing levels of investment per man. This is illustrated

TABLE 3.7 Changes in capital per man, U.K., U.S. and Norway

| *U.K.* Fixed capital per man, manufacturing | | | *Norway* Real capital per employee | | *U.S.* Horse power per production worker, manufacturing | | |
|---|---|---|---|---|---|---|---|
| | £(1970 prices) | 1920 = 100 | | 1920 = 100 | | No. | 1919 = 100 |
| | | | 1900 | 75 | 1899 | 218 | 65 |
| 1920 | 1756 | 100 | 1920 | 100 | 1919 | 333 | 100 |
| 1930 | 2200 | 125 | | | 1929 | 491 | 147 |
| 1938 | 2018 | 115 | 1939 | 137 | 1939 | 652 | 196 |
| 1948 | 2269 | 129 | | | 1954 | 958 | 288 |
| 1958 | 2753 | 157 | 1960 | 239 | 1962 | 1249 | 375 |
| 1968 | 3580 | 204 | 1969 | 331 | | | |
| 1974 | 4974 | 283 | | | | | |

SOURCE *The British Economy: Key Statistics, 1900–1970* (London and Cambridge Economic Service)
*National Income and Expenditure* (H.M.S.O., 1964–74)
*Monthly Bulletin of Statistics* (H.M.S.O., July 1975)
*Norway Statistical Year Book, 1974*
*I.L.O. Statistical Year Book, 1974*
*Income and Wealth Series VIII*, Goldsmith and Saunders (Cambridge, 1959)
*U.S. Annual Abstract of Statistics, 1973*

TABLE 3.8 Productivity and savings, 1970

| | *Average productivity of workforce: GDP/ total workforce* $ | *% of U.S. level* | *Savings as % of GDP* | *Savings per head of workforce* $ |
|---|---|---|---|---|
| *Developing countries* | | | | |
| Chile | 2835 | 24·2 | 17 | 482 |
| Argentina | 2783 | 23·7 | 20+ | 557 |
| Mexico | 2475 | 21·1 | 20 | 495 |
| Jamaica | 2186 | 18·6 | 25 | 547 |
| Algeria | 1493 | 12·7 | n.a. | n.a. |
| Brazil | 1369 | 11·7 | 16* | 219 |
| Zambia | 1351 | 11·5 | 27 | 365 |
| Sri Lanka | 494 | 4·2 | 21 | 104 |
| Bolivia | 423 | 3·6 | 16 | 68 |
| Nigeria | 411 | 3·5 | n.a. | n.a. |
| Thailand | 365 | 3·1 | 25 | 91 |
| Botswana | 306 | 2·6 | n.a. | n.a. |
| India | 294 | 2·5 | 16 | 47 |
| *Developed countries* | | | | |
| U.S. | 11,730 | 100 | 17 | 1994 |
| Australia | 6944 | 59·2 | 28 | 1944 |
| France | 6796 | 57·9 | 28 | 1903 |
| W. Germany | 6404 | 54·6 | 28 | 1793 |
| U.K. | 4625 | 39·4 | 19 | 879 |
| Japan | 3776 | 32·2 | 40 | 1510 |

* figures for 1969.

SOURCE *I.L.O. Statistical Yearbook 1974*
*U.N. Statistical Yearbook 1973*
*World Economic Survey 1973*
*U.N. Yearbook of National Accounts Statistics 1973*

by the figures above. Despite the manifold difficulties of measurement,[14] all indicators of investment or capital per head show increasing levels over time, in accordance with the need for ever-rising levels of labour productivity, and the rising level of savings per head.

In aggregate, total investment per head equals total savings per head, or

$$\frac{I}{L} \equiv \frac{S}{L} \qquad (1)$$

where

$$S = \text{total savings}$$
$$L = \text{the labour force}$$
$$I = \text{investment.}$$

Now

$$S \equiv s \, . \, O,$$

where

$$s = \text{average savings propensity}$$
$$O = \text{the level of output}$$

and

$$\frac{S}{L} \equiv s \, . \, \frac{O}{L}$$
$$\equiv s \, . \, o$$

where $o$ = output per man, or labour productivity.
Let $k$ = investment per head of the labour force.

Then (1) can be rewritten,

$$k \equiv s \, . \, o \qquad (2)$$

or investment per head of the workforce is equal to the savings propensity times average productivity of the workforce, In developed countries, the investment requirements of new techniques are such that equation (2) is realised without major distortions; in other words, over the long period new techniques are such that available savings per head roughly match available investment per head, at, broadly, full employment. Problems arise when advanced-country techniques are transferred unmodified to poorer countries. The transfer of technology involves transferring advanced country investment per head, or their $k$, to the underdeveloped economy. But available savings per head over the workforce as a whole are much lower, because of the lower labour productivity overall in the poorer countries. Table 3.8 compares labour productivity and savings per head in advanced and poor countries. With roughly proportionate savings functions, and overall per capita incomes in poor countries of some fraction, say one-tenth, of those in the advanced countries, savings per head, $s \, . \, o$, will also be one-tenth of those in advanced countries. If savings propensities are somewhat lower in poor countries, as suggested by the table, then the gap between savings per head is even greater than that between incomes per head. Thus while the transfer of technology involves transferring $k$ unmodified, $s \, . \, o$ in the underdeveloped country is much lower. But equation (2) is an identity which must be true. In realising the identity, major distortions occur.

In effect what happens is that all the investment resources have to be

TABLE 3.9 % of labour force that can be employed in F-sector using techniques designed for developed countries[1] (1970 values)

| | *U.K. technology* | *U.S. technology* | *Average of developed countries*[2] |
|---|---|---|---|
| Argentina | 63·4 | 27·9 | 33·3 |
| Jamaica | 62·2 | 27·4 | 32·7 |
| Mexico | 56·3 | 24·8 | 29·6 |
| Chile | 54·8 | 24·1 | 28·8 |
| Zambia | 41·5 | 18·3 | 21·8 |
| Brazil | 24·9 | 11·0 | 13·1 |
| Sri Lanka | 11·8 | 5·2 | 6·2 |
| Bolivia | 7·7 | 3·4 | 4·1 |
| Thailand | 1·0 | 0·5 | 0·5 |
| India | 0·5 | 0·2 | 0·3 |

[1] Assumed equal to ratio of savings per head in underdeveloped country to savings per head in advanced country.
[2] Average of six developed countries shown in Table 3.8.

SOURCE Table 3.8.

concentrated in the sector receiving the advanced-country technology, so that required investment levels per employee in this sector may be achieved. The size of this sector – the F-sector – is then limited by the ratio of total savings in the economy as a whole to the required investment/labour ratio in the F-sector. Thus if, for example, the advanced-country techniques require investment per head of ten times the level of savings per head in the economy as a whole, and if all the savings are used in the F-sector, only one-tenth of the labour force can be employed in that sector. The rest must be elsewhere.

Table 3.9 illustrates the consequences, for poor countries, of using techniques designed for advanced countries. The table shows the fraction of the labour force that could be employed in the advanced-technology sector, if techniques designed for the savings levels of advanced countries are adopted unmodified. Because savings per head are much lower, only a fraction of the labour force in India could be equipped with advanced-country technology – for example, only $\frac{1}{2}$ per cent of the labour force in India could be equipped with technology in line with U.K. resources, and only 0·2 per cent using technology designed for the U.S. The table is based on very crude assumptions: only one year's savings are considered, and it is assumed that no modifications at all are made to advanced country technology,[15] that there is a single technology (defined in terms of investment per man) for each country, so that selection of more labour-intensive techniques is not a possibility. While these assumptions strictly interpreted are incorrect, the figures are none the less illustrative of some of the consequences of transferring the most recent advanced countries' technology to poor countries.

The heavy investment demands imposed by advanced country technology thus concentrate countries' savings resources in that sector, to the neglect of the rest of the economy. The consequence is the emergence of a big and growing gap between the sector using that technology and the rest of the economy – between what we have described as the F- and L-sectors. The F-sector is associated with high labour productivity as compared with the L-sector. The L-sector suffers from lack of resources, and poor efficiency of the resources it does have as a result of the lack of scientific and technical research.

The big gap in labour productivity between the two sectors has a parallel in a gap between wage levels in the two sectors. There has been much debate on the causes of the high wages in the F-sector in most underdeveloped countries. in relation to the incomes earned elsewhere in the economy. It has been widely attributed to trade union and government intervention of various kinds.[16]

This is the institutional interpretation. However, the prevalence of the gap in many countries, despite institutional differences, suggests this is not the whole explanation. Moreover, institutions – trade unions and government wage policy – are not independent of the political economy of the country. Government policy is responsive to many pressures including that of the business community. The many links between governments and decision makers in industry suggest that governments would not support a 'high' wage policy, if it were not consistent with the interests of the industrial decision makers. Because of the massive potential supplies of labour from the L-sector trade unions have little effective independent power, unless they are encouraged by government and management. The relatively high wages must therefore be explained as being an outcome of the industrial system, rather than imposed institutionally from without.

It has been observed in developed countries that firms in an oligopolistic industrial structure do not keep wages at a minimum, but pass on part (roughly a constant share) of the gains in productivity to their workers.[17] This, broadly speaking, is what happens in poor countries too. But though this describes what happens it does not explain it. One explanation is that high wages in the F-sector are necessary for the efficient operation of the technology. Partly, this is in order to reduce labour turnover and maintain a stable work force.[18] In addition, required standards of health and hygiene, nutrition, punctuality, literacy and transport can only be achieved if wages are *considerably* higher than subsistence incomes in the L-sector. The more recent and sophisticated the technology the higher the standards of living required among the workers. Moreover, high wages are needed if the workforce is to consume the products it produces – with the sort of incomes obtainable in the subsistence sector, the work force would not provide a market for the products of advanced-country technology. The relatively high wages, therefore, in the foreign technology sector are in part at least the consequence of technological requirements; in so far as high (relatively) wages are necessary for the efficient operation of the technology, they are a characteristic of the advanced-country technology. While these relatively high wages are in large part a consequence of the type of technology adopted, they are also, again in part, a cause of the selection of this type of technology. The level of real wages is one of the factors determining the selection mechanisms leading to a particular choice of technique, described in Chapter 1. The existence of high and rising wages undoubtedly encourages the choice of more recent and capital-intensive techniques – in so far as a range of alternatives is available. But since these relatively high industrial wages are the *outcome* of the use of advanced-country technology, the interconnections between technology and wage levels represent one way in which the selection mechanisms are themselves the outcome of the technological choices made.

It can now be seen that a dualistic pattern of development is an inevitable feature of adopting rich-country technology in poor countries. Much of what has been described as the employment problem is in reality a syndrome of problems associated with the big gap in labour productivity and wages per head between the F-sector and the L-sector as discussed in the last chapter. Open unemployment results as workers attempt to secure the few jobs and high rewards of the F-sector, while underemployment, which is normally defined as relatively low productivity and incomes, stems from poor technology and underinvestment in the L-sector in relation to the F-sector.

The origin of these factors, leading to this duality, are all to be found in the characteristics of the advanced-country technology: concentration of investment resources in the F-sector, high labour productivity and high incomes (wages and salaries) in this sector in relation to the rest of the economy follow from the nature of the advanced country technology. Further, as will be discussed more below, the advanced-country technology produces high-income type products, ill-adapted to the needs of the poor in the rest of the economy.

*Skill requirements*: different techniques require different labour skills. The growth in incomes in developed countries has been accompanied by a vast growth in educational and training expenditure. Thus techniques designed in these countries have been able to assume high levels of worker skills and literacy at all levels. Table 3.10 shows how the composition of the labour force has changed in the U.S. during the past century. By 1974 only just over one-fifth of the U.S. labour force were unskilled, with corresponding implications for the skill requirements of U.S. technology. Managerial requirements have already been discussed. Countries which adopt advanced-country technology must provide similar skills, or import them. Thus the educational systems of the advanced countries have to be duplicated. But these systems are costly, and when introduced in poor countries only a minority can be so educated. The resulting educational system is designed to produce the sort of skills required by advanced-country technology. But it is totally inappropriate for education in the sort of skills required in the rural and self-employed sectors. For example, considerable attention is paid to educating students to high standards of fluency and literacy in some foreign language appropriate for the F-sector with its frequent foreign contacts. But for work in the rural sector and the L-sector where the majority spend their lives this is quite unnecessary. What is needed is a simple approach to accounting. What is taught is elaborate Western systems of accounting to the few privileged students who may become accountants in the F-sector. Similarly, with training for administration, for the law, and for engineering.[19]

TABLE 3.10 Composition of the U.S. Labour Force

| | *% of total* | | | | | | | |
|---|---|---|---|---|---|---|---|---|
| | *1910* | *1920* | *1930* | *1940* | *1950* | *1960* | *1968* | *1974* |
| Professional & technical | 4·4 | 5·0 | 6·1 | 6·5 | 7·5 | 11·2 | 11·2 | 14·6 |
| Proprietors | 23·0 | 22·3 | 19·9 | 17·8 | 10·8 | 10·6 | 9·9 | 10·4 |
| Clerks & kindred workers | 10·2 | 13·8 | 16·3 | 17·2 | 19·2 | 21·3 | 22·0 | 23·8 |
| Skilled & foremen | 11·7 | 13·5 | 12·9 | 11·7 | 12·9 | 12·8 | 13·2 | 13·3 |
| Semi-skilled | 14·7 | 16·1 | 16·4 | 21·0 | 20·4 | 17·9 | 18·4 | 16·1 |
| Unskilled | 36·0 | 29·4 | 28·4 | 25·9 | 29·2 | 26·2 | 25·3 | 21·7 |

SOURCE *Statistical Abstract of the United States, 1943* (Table 132); *1969* (Table 322); *1974* (Table 568).

The need for associated skills imposed by the use of advanced-country technology often leads to an apparently chronic shortage of skilled manpower. On the one hand, the shortage of local appropriately trained people appears to justify the development of the educational system along advanced-country lines, and encourages its further expansion. On the other, it requires the continued import of foreign personnel to carry out some of the skilled jobs. The shortage of local people and the need to import people from advanced countries together make for an escalation in salaries of skilled people, so that the differentials between skilled and unskilled tend to be higher than those in the advanced countries.[20] A particular case illustrates the position. In an interview about the possibility of Africanising technical posts, a foreign brewing firm said that the chief brewers needed a seven-year training in Edinburgh – until Africans had been through this training Africanisation would be impossible. Contrast these training requirements with those of the back-street or rural brewer to be found in every locality in Kenya. As so often, the advanced technology imposes advanced training requirements which are out of proportion with the training capacity of the country, and can only be realised by sending people abroad, or duplicating foreign training facilities in the country. The net effect is the disproportionate expenditure on education and training of the few in the F-sector, while neglecting the majority. In Kenya, for example, the cost per year of a student at the University is estimated to be in the region of £1000, (similar to costs per student in universities in developed countries), while the cost per student-year in primary school is estimated to be less than one-hundredth of this, as the table below shows.

TABLE 3.11 Annual costs per student (£K), 1968

| *Level of schooling* | *Operating costs* | *Capital costs* | *Total* |
|---|---|---|---|
| Primary | 7·7 | 0·1– 0·5 | 7·8– 8·2 |
| Lower secondary | 87·0 | 3·0–10·0 | 90·0–197·0 |
| Higher secondary | 175·0 | 10·0–14·0 | 185·0–189·0 |
| University | 690·0 | 625·0 | 1·315[a] |

[a] with increases in numbers, the costs per University student are declining and were estimated at £887 for 1971 by Fields (1974).

SOURCE Carnoy and Thias (1971), Table 6.

It is estimated that about half the children complete seven years of primary education. Of the remaining half, some receive no formal education at all, and others begin but do not complete primary education.[21] About 15 per cent of children go to secondary school, and about one-quarter[22] of these go on to further education. Table 3.12 estimates the proportion of total educational expenditure going to different categories of student, on the basis of these figures:

TABLE 3.12 The educational pyramid in Kenya

| *Level of education completed* | *% of population of relevant age* | *% of total expenditure on education* |
|---|---|---|
| None | 25 | 0 |
| Partial primary | 25 | 3·8 |
| Complete primary | 35 | 9·3 |
| Secondary | 11 | 32·3 |
| Further education —type I | 2 | 20·2 |
| Further education —type II | 2 | 34·4 |

SOURCE Carnoy and Thias (1971).
I.L.O. Kenya (1972).
Kinyanjui (1975).

NOTES It is assumed that partial primary school lasts four years and costs £8 p.a.; complete primary lasts seven years at the same cost; secondary education lasts four years costing £140 p.a.; further education, type I (non-university) lasts three years and costs £500 p.a.; further education type II (university) lasts three years and costs £1000 p.a.

Table 3.12 contains estimates rather than precise figures: it shows that 4 per cent of the relevant age-group consume nearly 55 per cent of total educational expenditure. These account for the upper echelons of employment in the advanced-technology sector, and for the teachers and civil servants who, on the whole, serve that sector. A further third of educational expenditure is devoted to those who stop education after secondary school – they account for much of skilled and clerical employment provided by the F-sector. The 50 per cent of children who do not complete primary school, and who are likely to spend their working lives in the L-sector, enjoy less than 4 per cent of total educational expenditure.

Moreover, it is not only a question of the quantity of resources but also of the content of education, as has been widely emphasised:[23]

> A more serious indictment of the educational system relates to its qualitative aspects. A growing number of people feel that the educational system is largely irrelevant to the great majority of the school-going population who do not reach the higher range of the educational ladder and cannot be absorbed in productive employment in the modern sector of the economy. Not only does the education they receive not equip them in any way for the kinds of life they will inevitably have to lead, but it is alleged to add to social and economic problems by the inculcation of inappropriate attitudes and values

> and by raising false expectations. All this leads to alienation and frustration of youth and a tragic waste of manpower. (Ghai, 1974)

The educational requirements imposed by advanced technology are not just a matter of specific training – but also of the whole environment in which children in advanced countries are brought up. This environment – the style of life as well as its formal content – need to be largely duplicated.

Technology imposes requirements on mental processes too: the requirements of modern technology, both productive technology and the bureaucracy associated with it, are far removed from those of traditional patterns of production. The consequences for economic efficiency and mental equilibrium if the consciousness of traditional society is maintained side-by-side with advanced country productive processes, are described by Berger *et al.* in *The Homeless Mind* (1973).

Alternative life styles are closely associated with patterns of demand for consumer goods. In requiring a life-style of its managers, engineers and workers approaching that of advanced countries, the technology also leads to patterns of demand related to those found in advanced countries. The most direct connection of this type can be seen in demand for clocks and watches, buses, cars, and bicycles, typewriters and biros which are essential to the efficient operation of the technology. Indirectly, demand for television and detergents, for mechanical toys and processed food are aspects of an advanced technology life-style.

The adoption of advanced-country technology on any scale thus affects patterns of demand, from the point of view of securing its efficient operation. Techniques also have more direct implications for demand.

Each technique, as argued in Chapter 1, is associated with a particular product, which has a set of characteristics, consisting of use, style, material, quality, etc. As incomes rise the characteristics of the products consumed to fulfil various broad functions change – indeed this change in product characteristics to a large extent *constitutes* the increase in real income. Techniques designed for consumption in a particular society will therefore also be designed to fit in with the income levels typical in that society. When techniques designed for use in rich countries are transferred to much poorer countries, the products produced by those techniques are transferred too. The two – the techniques and the products – are inseparable aspects of the technology. Thus products designed for consumption in much richer societies are transferred to economies where, on average, incomes are much lower. Two effects may ensue. Unequal income distribution creates a market for the high-income products among the elite. Alternatively (or additionally), imbalanced consumption patterns permit consumers with typical income levels to consume a few high-income products and little else. The situation may be illustrated by taking an actual product – say a car. A car (historically and within nations at any one time) is a product associated with rising incomes. Suppose car production technology is transferred to poor countries. Let us assume that the car is relatively cheap – say costing about one-third of annual earnings in the U.K. If car production and consumption is transferred to a third world country with e.g. one-tenth of the income of the U.K. the cost of the car would represent over three times annual earnings. If incomes were sufficiently unequally distributed then a small number of consumers in the third world country might secure incomes high enough to be able to buy a car without a complete imbalance in their consumption patterns. But then the rest of the population would have to be that much poorer to permit such income levels for the elite. Alternatively, if workers spent *all* their income for three years, they could purchase a car, but this would imbalance their consumption so that their other needs – for food, shelter, entertainment – were totally (and impossibly) neglected. These are two stark alternatives. In practice a bit of both occurs: unequal incomes create markets for the high-income products, while imbalanced consumption patterns extend these markets. The latter – imbalanced consumption patterns – occurs particularly in relatively low-income products, such as processed foods, bottled drinks, cigarettes. These products (unlike the car) are sufficiently inexpensive to be possibilities for fairly low income earners, but by consuming them the consumers tend to imbalance their consumption, spending too much on such goods and having insufficient over for the other products and other needs. Each of these goods – however inexpensive in itself – contains the standards designed for higher income levels in the advanced countries and tends to be inappropriate for poorer consumers.

It has been argued that the inappropriate products argument is inapplicable to the sort of cheap product which is widely consumed, as those listed above, and applies mainly to consumer durables, etc. While the significance of the imbalance is clearest in the case of consumer durables even widely consumed products may contain excessive standards in relation to the income levels of consumers and the needs fulfilled. Looked at in another way, it is clear that if one started *ab initio* to design products suitable for consumption by very poor consumers the products would differ in function and quality and packaging from those transferred from richer (and different) societies.

This is shown by the nature of the products produced in the subsistence sector and informal sector – where the products are designed for own-consumption and sale to poor consumers. Shoes made of used car tyres, simple furniture, low cost housing well below the standards of any produced in the 'modern' sector, locally brewed beer, etc., are all of a completely different quality from those available in the foreign-technology sector. These goods, though often ill-designed and of low efficiency, indicate the sort of characteristics of 'appropriate' products. Similarly, traditional products like soap, processed food and bricks are examples of goods designed in and for consumers in third world countries. They contrast with the goods with similar functions produced by advanced technology, which are far more resource-consuming than the traditional products.

The point may be illustrated diagrammatically.

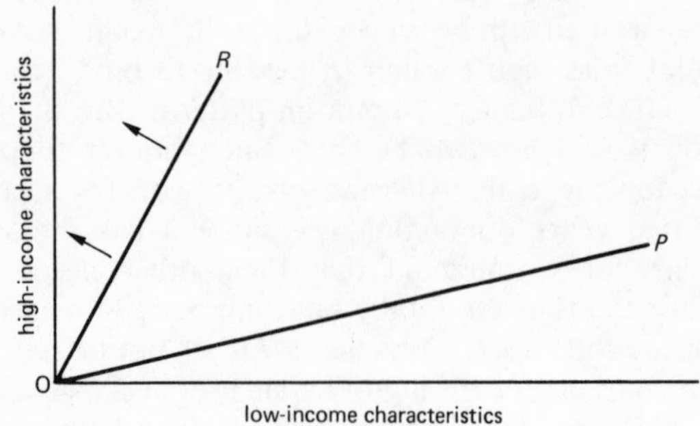


FIG. 3.1

In the diagram ray *OP* shows products appropriate to a poor society, while line *OR* shows the sort of products appropriate to rich societies. Over time in advanced countries, as incomes rise, there is a tendency for the product characteristics to shift up and to the left, as shown by the arrows, as later products are designed for higher-income consumers. If the new products simply *supplemented*, without affecting, the old products, this might not have serious implications for the poor societies, apart from the way in which their consumption patterns were changed towards the less appropriate products via the demonstration effect and advertising expenditure. But the new products tend to replace the old products, or increase their cost, for two reasons. First, because of the existence of economies of scale which mean that it is often inefficient to maintain production of both sets of products when both are fulfilling broadly the same needs. Secondly, because technical progress is concentrated on the new (rich-country) processes and products, which tends to mean, as argued in Chapter 1, that the new products are more efficient than the old – i.e. a given amount of resources gives a greater amount of need-fulfilment. Hence the old products, represented by *OP*, which would be more suitable for consumption in poor countries, are eliminated and only the new high-income products are available. Put in more concrete terms, the steam engine is rendered obsolete by diesel or electric ones.

While the transfer of inappropriate products requires inequality of income distribution if adequate markets are to be created, the transfer also contributes to the required inequality. In part this is a matter of factor use and factor rewards flowing from the technology. High wages, relative to incomes in the economy as a whole, are generated by the advanced-country techniques, as discussed above. The technology itself imposes certain requirements for the consumption levels and patterns of workers – these requirements, for example for standardised and Western style dress, means of transport, wealth and education both lead to inequality in the society and create a market for the advanced-country products the technology produces. More extreme sources of inequality arise from the skill differentials, which again, as argued above, stem from the technological requirements. Once again the high incomes (relatively) so created among the elite provide a market for the inappropriate products and so appear to justify the technology. But this is because the technology is self-justifying, creating its own demand.

In the case of public policy, particularly public goods, there is a direct connection between the nature of the product produced and the distribution of income. The provision of particular types of public good itself directly affects the distribution of income. Given a limit in total public expenditure the allocation of funds to high-income inappropriate products automatically removes resources from low income consumers. For example, the provision of tarmacked highroads is at the expense of improving roads for the rural areas. Similarly with education. This is a matter of big lump sum expenditures on items which serve the richer members of the community and in so doing further enrich them and impoverish others. It is also a matter of excessive standards on small items – like e.g. books and blackboards, buildings, hospital beds – all of which involve allocating more resources to these items, and therefore to their consumers, than is consistent with a more egalitarian and balanced allocation of expenditure. In such cases, the transfer of inappropriate goods directly contributes to inequalities – though it is also encouraged by the existing inequalities, with pressures from the elite for high-income advanced-country-type products.

Government policy creates a market for inappropriate products in other indirect ways: the minimum standards laid down by many governments for housing, food processing, etc., often copied from the standards in advanced countries, often make appropriate products illegal, thus preserving a market for inappropriate products.

### LINKAGES AND 'TECHNICAL FACTORS'

As argued in Chapter 1 there are all sorts of links between different techniques, so that a particular technique imposes requirements on the rest of the system, and will only be viable if these requirements are met, within certain limits. Sometimes the limits are wide, or considerable variation is possible: in other cases the technique requires inputs of a particular quality and specification such that there is very little choice in associated production methods. It is on the whole a characteristic of more recently developed techniques that the room for variation in associated techniques has declined. This has occurred partly as a result of the more sophisticated scientific techniques involved, so exact temperature control, etc., is required, and also with mechanisation and then automation. Mechanical methods impose more rigorous requirements than hand methods because hand methods may correct for variations in previous processes. Chapter 10 below illustrates this for cement block manufacture. Automated techniques leave even less room for human correction of previous variability – and hence impose strict requirements of standardisation and detailed specification on previous processes. The effect is that a given automated technique is designed to be operated using the output of certain specified techniques for previous and subsequent stages, and variation may not be possible. Thus the links between different parts of the system have got stronger. Contrast, for example, variation possible on a nineteenth-century steam engine with that in a nuclear plant; or the variation possible in the parts of a horse and cart with that in a car from an automated assembly line.

The links between techniques extend to requirements for particular types of transport (and therefore roads), of legal and administrative services and so on, as well as for particular products as inputs. Managers are trained to work with a particular set of linked services – these include secretarial facilities, postal services, banking, insurance, law, etc. In so far as importing technology leads to importing managers, or at least managerial techniques, as argued earlier, this leads to requirements for the linked services over and above those requirements which are strictly related to the hardware technology.

All these links mean that the introduction of a single technique from the advanced countries imposes requirements for linked techniques in the rest of the economy. Where the rest of the economy fails to produce the required linked processes, the original enterprise may produce them itself, or put pressure on the government to obtain their production, or, where possible, import the linked process.

Techniques designed in a particular country are designed to fit in with the particular technology package in use in the system. Hence the linked requirements are automatically met. But where a technique is imported into a different environment, linking processes are not available. Then, if the techniques are to be efficient, the linked processes must somehow materialise: in other words the technology package which the initial technique was designed for must be reproduced. Hence the input of technology leads to new requirements for imports, or imported techniques (which in turn bring their own linked requirements) and for the type of infrastructural services found in the initiating country. These requirements are often totally alien to local systems of production of goods and services. These then tend to become irrelevant, starved of resources, while the country concentrates on provision of the processes linked to the advanced-country technology.

We have considered the nature of techniques under three headings – organisation of production; income levels and the implications for resource availability, resource use and patterns of consumption; and the technological system. Each, in a way, leads to similar conclusions: individual techniques are designed for a particular economic/technical environment, and are efficient, indeed viable, often only in the context of that environment. If they are transferred to a completely different environment – from advanced country to underdeveloped country – then the original environment has to be reproduced. The attempt to achieve this reproduction has three effects: first, it involves heavy reliance on foreign sources for personnel, for materials, parts and servicing, for technology for linked processes, and often for markets. Secondly, it requires concentration of the underdeveloped country's resources on the minority F-sector to provide the resources and demand that the technology requires to duplicate advanced country conditions in the poor country, and thus alleviate the total dependence on foreign sources which otherwise and initially occurs. Associated with this concentration of resources in the F-sector goes the neglect of the rest of the economy, which starved of resources appears increasingly feeble and uncompetitive, a source of low productivity and poverty and not much else. Thirdly, because the requirements imposed by the technology are alien to local history and resources the production achieved tends to be inefficient, compared with the operation of similar techniques in the advanced countries for which they were designed.[24] The reproduction is imperfect: many of the criticisms of poor countries – for inefficiency, laziness and corruption – are due to this.

TRENDS OVER TIME

The situation is not static. Technology changes all the time in advanced countries, while economic conditions there also change in parallel with the technical changes. Some of the earlier tables showed aspects of this – the growth in capital per man over time, increasing wages per man, and skills. These trends can be expected to continue with new (higher-income) products replacing old as incomes rise in advanced countries, greater capital per man of new equipment, higher skill and scale requirements. With continued concentration of resources on rich-country technology, its relative efficiency as compared with indigenous methods is likely to increase. The technology is thus likely to get increasingly out of line with poor countries' needs – or increasingly inappropriate – while simultaneously appearing increasingly unavoidable, as the only efficient technology available.

The way in which technology develops over time may be illustrated in the case of textiles. A study of choice of techniques for Latin America compared three technologies, a 1950 vintage, a 1960 vintage and a 1965 vintage. Table 3.13 shows how some of the key variables changed with technological developments between these years.

TABLE 3.13

| 1950 = 100 | *1950 vintage* | *1960 vintage* | *1965 vintage* |
|---|---|---|---|
| Investment per unit of equipment of minimum economic size | 100 | 127 | 146 |
| Investment per worker | 100 | 190 | 310 |
| Value added per worker | 100 | 145 | 211 |
| Output per $ of investment | 100 | 76 | 68 |

SOURCE ECLA Report, Table 17 (1966).

Scale, capital per employee and labour productivity increased during this period. In this case the capital requirements per unit of output also increased, though rather little between the 1960 and 1965 vintages. Changes in the skill composition of the labour force were also marked, as Table 3.14 shows.

TABLE 3.14 Labour requirements in textile industry

| | *Conventional* | *Intermediate* | *Automated* |
|---|---|---|---|
| | | *As % of total requirements* | |
| Administrative | 5·6 | 6·9 | 8·1 |
| Skilled | 24·5 | 31·5 | 32·3 |
| Semi-skilled | 41·9 | 41·5 | 41·4 |
| Unskilled | 28·0 | 20·0 | 18·2 |
| *Total* | 100 | 100 | 100 |
| Total requirements per unit equipment | 100 | 76·6 | 58·4 |
| Total requirements per $ investment | 100 | 63·7 | 31·9 |

SOURCE UNIDO Report (1968). The conventional/intermediate/automated classification roughly corresponds to the vintage classification used by the ECLA Report.

Labour requirements fell at all levels, but the fall was far more marked in the case of unskilled labour (the automated machine required only 38 per cent of the requirements of the conventional machine) while the proportionate importance of administrative and skilled labour increased.

At a macro level it has been estimated[25] that technical change was such that a ten-year technology 'freeze' in Puerto Rico would have increased employment by 40 per cent in 1963. Baer and Hervé (1966) (for some Latin American countries) and Gouverneur (1971) (for the Belgian Congo) show how capital per head has increased over time in underdeveloped countries, paralleling changes in the developed countries.

The use of the term 'inappropriate technology' is shorthand to describe the many respects in which advanced-country technology tends to be inappropriate, as discussed earlier. Of course, not *all* technology is necessarily inappropriate in *all* respects. Theoretically, it is possible that a single technique may use inappropriate methods of production to produce appropriate products, and conversely. Much has been made of this possibility by Helleiner (1975):

> Just as there is no presumption that capital goods industries will be capital-intensive there is no presumption that 'appropriate' products will employ 'appropriate' [i.e. in this context, labour-intensive] technology.

However, because technological developments occur in a historical context there tend to be links between different characteristics, such that a technique which is inappropriate in one respect is very often inappropriate in other respects too. For example, techniques designed for use by high-wage labour (and therefore labour-saving) tend also to be designed for consumption by that same high-wage labour (and are therefore high-income products). At the other end of the scale, techniques designed in and for the subsistence sector use the resources available there – and are therefore labour-using – and produce products for own consumption, i.e. low-income products.[26] In the subsistence sector, one could argue that the relationship between appropriate methods and appropriate products is a logical one. Put in Marxist terms, since in this sector production is for use value and not exchange value, the products must be appropriate. It is where exchange takes place that a divorce is possible between the nature of production methods and what is produced. Here the relationship between characteristics has historical origins. In so far as the techniques are designed for production and consumption in a particular society, then the production and consumption characteristics of that technology will reflect conditions in that society – if it is high-income then both will tend to be suitable for high-income societies, and inappropriate to low-income societies. However, with unequal income distribution it is likely that some techniques designed in low-income societies are intended for the rich in those societies. Consequently, some labour-using methods will tend to be associated with high-income products. This is clearly true of many of the products – such as palaces and pyramids, silks and satins, rickshaws and servants – which were developed in traditional societies. Indeed, in very poor societies, where the surplus available for consumption on anything but the barest essentials of food and shelter is very limited, the development of such non-essential products can only occur in the context of inequality. The agricultural and industrial revolution after a time permitted mass incomes to rise above the minimum food/shelter level, and thus to provide a market for manufactured products among average-income consumers. In modern industrial societies, the mass market (the income at the mode) may provide the best long-run prospects for new products, but many products are initially designed for relatively high-income consumers, for these consumers are prepared to pay the high cost of the initial experimentation; as scale of production expands, teething problems are ironed out and the products reach the mass market, whose income per capita meanwhile has risen towards the level of the high-income consumers to whom the product was initially sold. This is an aspect of Vernon's product cycle[27] on the consumption side. At the initial high-cost stage, products are intended for above-average-income earners; at the mass production stage they are intended for the mode. Over time incomes at all levels generally are increasing, so that each new vintage of products is designed for higher-income consumers than the vintage before. According to this model, it is unlikely that products will be designed for below-average-income consumers – because in a dynamic context their absolute incomes are always expected to rise, so that they grow into the soon-to-be rejected products of the income classes above them. Just as in a family it is rational to buy or make new clothes for the older children, who can hand them down, not for the younger who cannot, so product development in a growing economy tends to keep ahead, rather than lag behind, average incomes. Similar arguments apply on the resource side; with labour productivity and real wages expected to rise over time, a firm introducing a new technique will design it to allow for this, so that it does not become obsolete as soon as it is in use, and will therefore design new techniques to produce with above the current average degree of labour saving. New techniques will tend to exhibit both high labour-saving characteristics, and high-income products, relative to the average in the society in question. Thus, although there may be exceptions, there will be some tendency for techniques from advanced countries to be inappropriate simultaneously in a number of respects.[28]

One important exception would occur if techniques were designed for production in one society, at one income level, for consumption in another, at another income level. On the one hand, advanced societies might design techniques to sell products purely for export to poor societies; conversely, techniques might be designed for operation in a labour-intensive way in poorer societies for export to high-income societies. On the whole, these possibilities represent theoretical exceptions rather than likelihoods because exporting tends, as powerfully argued by Burenstam Linder (1961), to follow success in domestic markets. But the recent growth in the location of labour-intensive parts of the productive process in LDCs by multinationals[29] gives some reality to this possibility.

To sum up this part of the discussion, we may conclude that in the subsistence sector there is a necessary connection between characteristics of production and consumption; in the monetary sector the connection is not necessary but depends on the historical development of the technology; in traditional, static societies, there is a strong tendency for labour-intensive methods to be used to provide high-income products for the rich; in modern dynamic industrial societies, there is likely to be a positive relationship between higher-income products and more labour-saving (per unit of output) and more capital-using (per unit of labour) methods; however, an exception to this is possible for those techniques designed primarily for trade between rich and poor (or poor and rich) countries.

It may therefore not be possible to classify any particular technique as inappropriate or appropriate, because it may be appropriate in some respects, inappropriate in others.

### SELECTION MECHANISMS

It was suggested earlier that within the limited choice available more recent and capital-intensive techniques are often selected than is strictly necessary. This was attributed to the fact that the selection mechanisms are themselves part of the technology system, so that the system becomes self-justifying, generating selection mechanisms which are consistent with it. We can now be more explicit about this conclusion.

In Chapter 1, it was argued that choice of techniques from within those available depends on the interaction of the goals of decision makers and the characteristics of the techniques, given the constraints facing the decision makers. Different groups of decision makers have different goals and face different constraints. The actual decision therefore varies according to the nature of the decision maker; for the economy as a whole the net effect depends on the balance of control over resources of different types of decision maker.

The existence of an F-sector using advanced country techniques tends to bias additional choice of technique decisions towards the selection of more recent and capital-intensive techniques because it enhances the relative advantage of such techniques; it leads to factor prices favouring such techniques, and it gives greater control of resources to decision makers who are likely to favour such techniques. We shall consider these points in turn.

Linkages between different techniques occur in terms of the specifications of associated products and services, such that any technique requires particular types of input and subsequent processing. Later techniques from advanced countries are designed to be operated with other techniques and products of the same vintage. Thus once an F-sector has been established it creates demands for other products, and traditional products and techniques are put at a disadvantage in serving the F-sector. Similarly, international demand is oriented towards the later products in a linked fashion.[30] Consumer demand is also shifted towards later products and techniques. This is in part a matter of advertising effort, which is concentrated on the later F-sector products. It is also a product of the income differentials generated by the F-sector, with relatively high salaries among the managerial and skilled workers, and relatively high wages among the unskilled factory workers. Links between the different consumption products – e.g. radios and batteries – also play a part. Thus the F-sector directly, through its demand for inputs, and indirectly via its effect on consumption patterns, creates a demand for other products of similar vintage. But a demand for a particular product is a demand for a particular technique. Older and more labour-intensive techniques find it difficult to compete with newer techniques because the older products they produce are not in demand by the producers and consumers of the F-sector.

The older products – in so far as they remain efficient – may still satisfy consumers in the informal and traditional sectors, as may their own products. Indeed the L-sector can only make effective use of such products. Later products are normally too expensive for them, embodying excessive characteristics. But the demand of this sector is small in relation to the demands of the F-sector, because of the low-incomes of this sector. These low incomes derive from the low investment in the sector and the low productivity of the investment, in the absence of an effective technology.

The economic system itself generates the factor prices and factor rewards which form the incentive for further economic activity and help determine technological choice. The F-sector, as has been suggested above, tends to lead to a system of factor rewards and factor availability radically different from those obtaining in the L-sector. Because of the international nature of the capital and technology markets and much of the market for the higher level skills used in the F-sector, both earnings and availability of these factors are largely internationally determined. Firms using advanced techniques have access to international capital and labour markets, so that it is international availability of these factors and international prices which tend to be decisive rather than local conditions. This, of course, is peculiarly true of multinational corporations, but it is also true to a lesser extent of large local firms using foreign technology and often hiring foreign managers. With the growth in joint ventures of various kinds many of the characteristics of multinational firms are now shared by local firms with strong foreign interests. Hence, in the F-sector, the price of additional finance tends to be low in relation to conditions elsewhere in the economy and it tends to be relatively easy to secure. The price of management and skills tends to be high, in relation to rewards elsewhere in the economy, pulled up by the high pay in the advanced countries. The F-sector also tends to pay its unskilled labour force considerably more than alternative occupations. This, as has been argued, is both cause and consequence of its high levels of labour productivity. The net result is that the relative price of different factors in the F-sector tends to be different from that in the traditional parts of the economy. Finance costs less (and is more readily available) and labour costs more, so that a more capital-intensive selection of techniques is encouraged, than if the price/availability ruling in the L-sector were operative. The diagram illustrates how this might lead to the choice of a more investment-intensive and more recently developed technique:

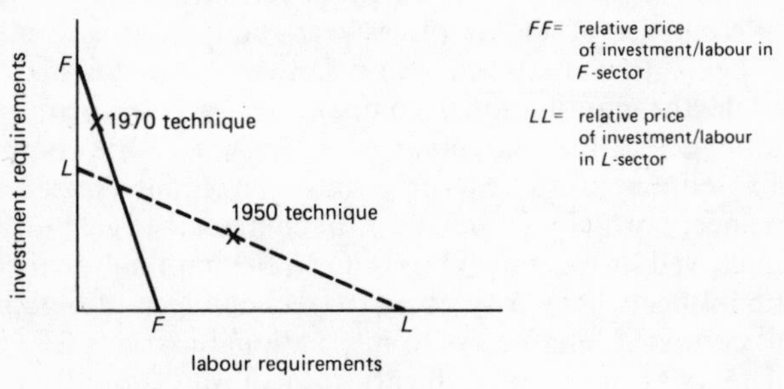


FIG. 3.2

As the diagram shows, with prices, *LL*, in the L-sector the more labour-intensive 1950 technique is most profitable, But with the prices ruling in the F-sector the more investment-using 1970 technique is most profitable.

The diagram suggests that if L-sector prices could be introduced into the F-sector then a more appropriate technique would ensue, being more labour-using and less investment-using, in accordance with the true factor availability in developing countries. It is sometimes argued that the local sector prices truly reflect the factor endowment of the under-developed countries and that the F-sector prices are 'distortions', institutionally induced. However, the analysis here suggests that while such prices may in one sense be 'distortions' leading to an over-invest-ment-using choice of technique, they are themselves the outcome of the use of advanced-country technology; by locking the sector into the advanced-country technological system that part of the economy also gets locked into international prices of capital and skilled labour, and relatively high wages for unskilled labour. Hence the prices cannot be effectively corrected for 'distortions' (or brought nearer to those ruling in the L-sector) until the technology in use is changed away from recent technology from advanced countries, and towards more appropriate technology. Put in another way, the ruling prices are not parameters that may be varied by the government – they are themselves the outcome of the economic system. The institutions – trade unions, banks, wages councils and the governments – which are credited (or debited) with distorting the prices, are, on this view, simply the instruments bringing about a system of prices that is consistent with the technology in use.

The third way in which the technological system is self-justifying is in determining the balance of resources controlled by different types of decision maker. For many reasons – some of which were discussed in Chapter 1 – technological decisions made by subsidiaries of multinational companies are likely to differ, in a systematic way, from decisions made by local small self-employed enterprises. The former face a different set of factor prices/availability than the latter – the difference being that discussed above between F-sector and L-sector prices. While the multinational company has access to more or less unlimited finance at the marginal cost of finance to the company as a whole, the self-employed enterprise has very limited possibilities of raising finance, normally at high cost. In contrast, the cost of labour to the self-employed firm is much less than to the multinational company. Multinational firms have easy access to technological developments in the parent company, and access to international markets for input and output. The small firm has difficulty in acquiring recently developed technology and relies largely on machinery purchases for acquisition of technology. The small firm's market is limited to the local market. The objectives of the two also differ: the self-employed firm may wish to maximise total income (of himself and his family employees), while the multinational enterprise is likely to maximise worldwide profits. The net effect of these differences is that the multinational firm tends to operate in the world system serving international markets, using large-scale plant, producing the more recent products, and adopting some of the latest techniques, whereas the self-employed firm operates on a small scale, uses old equipment, produces older products, and caters for the low-income local market. Other types of enterprise tend to come somewhere between these two. An empirical example of these differences is contained in the discussion of maize grinding in Chapter 9. The net effect on technological choice for the economy as a whole depends on the balance of resources controlled by different types of decision maker.

The existence of a sector using advanced-country techniques – the F-sector – gives a competitive edge to similar advanced-country techniques, since, as argued above, there are links between advanced-country products. This in itself increases the proportion of investment accounted for by the large-scale sector firms (local or foreign) because the small often cannot produce the type of product required. Moreover, the pricing system tends to be such that more resources are accumulated in the F-sector, relative to the L-sector. The F-sector is fundamentally oligopolistic, while the L-sector is competitive. The terms of trade may therefore favour the F-sector, which can charge high prices relative to those charged in the L-sector. This tendency has been discussed for the manufacturing/agriculture terms of trade, but the discussion also applies to F-sector manufacturing versus L-sector manufacturing. The (relatively) high prices in the F-sector consist of relatively high payment to all types of labour and relatively high profits. The profits are available for ploughing back into the firm, so increasing the control of resources by this type of firm. The high salaries and incomes are, to a much larger extent than L-sector incomes, spent on further F-sector products and so also return to the sector. Consequently, there is a cumulative increase in the resource control of large-scale firms as compared with the self-employed enterprises. Moreover, the F-sector firms may also invest from resources borrowed internationally, while the L-sector is confined to its own resources. A further source of finance – the government and local banks – tend to favour the F-sector.

Large-scale firms, both local and foreign, also tend to gain control over the government machine, so that government policies and government institutions favour the F-sector in relation to the L-sector, so encouraging relatively faster accumulation in that sector. For example, credit institutions and banks are oriented towards the large-scale foreign technology sector. Government policies on standards and product specifications, protection and employment, are all primarily directed at building up the F-sector. Policies which protect the F-sector often directly harm the L-sector – since they increase the cost of any resources they acquire from the F-sector, and make it more difficult for them to establish markets. The prevalence of government policies favouring the F-sector is partly due to the belief that industrialisation using foreign technology offers the only way of escaping underdevelopment. Also, members of the government acquire direct interest in the fortunes of firms in the F-sector.

The existence of a sector using recent techniques from the advanced countries thus tends to bias the selection mechanisms in favour of the more recent techniques from advanced countries. Thus where there *is* a choice of efficient techniques, the more recent technique may be selected. Given the context in which the decision is made, the more recently developed technique is the most profitable from a private point of view.

Techniques of social cost/benefit analysis offer partial correction, leading sometimes to the selection of older and more labour-intensive techniques. But their correction is only partial – for they correct the relative prices (of labour and capital, of production for home consumption and for export), but they do not change other aspects which are often of much greater importance. In particular, they very often take as parameters product specification, the nature of the decision maker, the scale of production and the type of market for which the product is intended. But, as we have seen, these variables are often of decisive importance in choice of technique, of greater significance than relative prices. In one way it is right that social cost/benefit techniques do treat these variables as parameters for in the context of a single decision they are parameters. Only if the whole technological system is called into question do these parameters become variables. Social cost/benefit analysis which is aimed at single micro-decisions is not suited to assess systems as a whole.

The endorsement of the choice of recent techniques by the criterion of private profitability, even where there is a degree of technical choice, suggests that what is at fault is not the technological choice, but the selection mechanisms. After all, if where more labour-intensive techniques are available they are rejected, there seems little point in developing additional techniques, which would presumably also be rejected. What is needed then, it is argued, is reform of the selection mechanisms. This conclusion is supported by social cost/benefit analysis in so far as, by choosing among techniques with different and 'socially' determined relative prices, a more labour-intensive alternative is selected than that chosen using actual prices. Hence, the concentration on changing relative prices, directly via changes in taxes/wages, etc. or indirectly via social cost/benefit analysis. But the analysis above suggests this exclusive concentration on relative prices is seriously misguided for two reasons. First, because relative prices are only one, possibly minor, determinant of technical choice. Secondly, because relative prices and many of the other determinants of technical choice *are not independent* variables within the control of the government but are themselves the outcome of the technological system adopted. A change in technology is thus a requirement for a change in relative prices and of many of the other determinants of choice. Of course, causation does run both ways – it is impossible to change the technological system without some change in the selection mechanisms, by definition. But the critical change required may not be that of relative prices, but the distribution of income and the resource control of different decision makers. These too, as argued above, are in large part determined by the technological system in use. Whether, therefore, it is possible to make a break to an alternative system is a very difficult question, which will be discussed more in the next chapter.

NOTES

1. In consumer durable expenditure for example, the switch to super-8 cameras has necessitated a parallel switch in films and projectors.
2. See Sercovich (1974) and Langdon (1975).
3. Both the neo-classicists' concentration on prices and the radicals' concentration on political power are examples of exclusive emphasis on changing the selection mechanisms, rather than seeking alternative technologies.
4. See Kay (1974) for a comparison of the medieval system with that in Latin America, and for references to many other studies.
5. See e.g. Worley (1962).
6. See, for example, Bain (1956), Chenery (1953), Merhav (1968), United Nations Bureau of Economic Affairs (1959) and (1964), Pratten (1971).
7. The estimates are for production technology: they ignore marketing and transport costs, so that the most profitable technique in any particular situation need not be at the minimum efficient scale as defined by Pratten.
8. Quoted by Winston (1971).
9. Little, Scitovsky and Scott (1970), Chapter 3, summarise some of the evidence on capacity utilisation in underdeveloped countries.
10. See e.g. the description of local/traditional forms of organisation in Marris and Somerset (1971), and Shetty (1963).
11. Notably McClelland – see McClelland and Winter (1969).
12. Some of the explanation of the (possibly unjustified) capital-intensive choice of techniques found in the Strathclyde Study (see Pickett *et al.* (1974)) and the differences in technology of multinational firms as against local firms found by e.g. Wells (1973) may lie in the specialist advantage MNCs have in making use of recent advanced-country techniques. See evidence and discussion in Morley and Smith (1974).
13. For examples see the Bulletin of the Intermediate Technology Development Group, *passim.*
14. See the discussion in Chapter 8.
15. Apart from the possibility of modifying the 'core' technology, it is well established that there are many opportunities for using more labour-intensive methods in ancillary activities – see Pack (1974), and Ranis (1972(a) and (b)).
16. See e.g. Smith (ed.) (1969), Turner's advice to the Tanzanian Government (I.L.O. 1967), Harris and Todaro (1969).
17. See Sylos-Labini (1969).
18. This was the reason why the Carpenter Report (1954) in East Africa recommended a rise in African wages.
19. See, for example, the discussion of engineering training in Kenya, in the I.L.O. Employment Mission's Report (1972).
20. Hinchcliffe (1975) shows the ratio of gross discounted lifetime earnings (at a 10 per cent discount rate) of those with higher education to those with primary education as follows:

| | | | |
|---|---|---|---|
| U.S. | 2·8 | Philippines | 2·6 |
| Canada | 2·9 | Ghana | 9·1 |
| Israel | 2·6 | Kenya | 7·6 |
| Mexico | 3·8 | Nigeria | 10·3 |
| Colombia | 5·2 | India | 5·7 |

21. I.L.O. Kenya (1972), p. 238.
22. This figure is derived from the survey of school leavers reported by K. Kinyanjui (1975).
23. See I.L.O. Kenya (1972).
24. For example, a survey of foreign investment in Latin America found that Latin

American industry which used sophisticated techniques was relatively inefficient as compared with similar operations in advanced countries – see U.N. Department of Economic Affairs (1971).

25. By Weisskoff *et al.* (1973).
26. James (1976) shows how this operates for handloom textiles in India, so that the most labour-intensive methods also produce products consumed by the poorest consumers. Ganesan (1975) shows that labour-intensity of building methods, in Sri Lanka, is greater the lower-income the type of house.
27. See Vernon (1966) and Hirsch (1967).
28. Here we have only considered two characteristics. Similar arguments apply to other characteristics.
29. Described by Helleiner (1973).
30. For more discussion of the international dimension, see Chapter 7.

# 4 Appropriate Technology

There is ambiguity as to what counts as appropriate technology. According to Morawetz (1974), 'Appropriate technology may be defined as the set of techniques which make optimum use of available resources in a given environment. For each process or project, it is the technology which maximises social welfare if factor prices are shadow priced.'

This definition takes as given the set of techniques available, and defines appropriate technology as the best choice within the available set, using shadow prices to select that best choice. The definition is open to two objections. First, it implies that society may arrive at a unique set of social shadow prices to select the optimum technique. In reality conflicts between different parts of any given society mean that different groups have different objectives, which discredits the concept of a single set of shadow prices and a single optimum.[1] Secondly, and particularly important in relation to discussion of technology, is the mistaken assumption of a given set of techniques. The main point of the discussion of appropriate/inappropriate technology is thereby missed; as argued earlier, the whole thrust of technological development has been such as to create an entire set of inappopriate techniques, and to leave undeveloped and underdeveloped the techniques which suit the conditions in poor countries. There are two issues here: one is the question of choice within existing techniques, or what we might describe as the 'appropriate choice', the other is the question of the development of more appropriate techniques. It is this latter aspect, implicitly ignored in the Morawetz definition, which is the main concern here. Once one takes as a potential variable the set of techniques developed, it becomes impossible to define any particular set as being 'the appropriate technology'. All that can be said is that some techniques are more appropriate (or less inappropriate) than others. The point is that it may always be possible to improve upon any existing set of techniques, making them for example, more efficient, and hence one cannot say that one has arrived at *the* appropriate technology.

A further difficulty arises from the argument just discussed – that given conflicting groups within society, their interests and objectives may vary, and so therefore will the sort of techniques each group considers appropriate. In much of the earlier discussion, there has been, implicitly, the sort of assumption criticised here, that it is meaningful to talk of *social* goals and therefore of what is appropriate to society as a whole. This is only so to the extent that the interests of society as a whole are identified with those of particular groups. A move to the sort of technology described as appropriate here would, for the most part, reduce the relative and probably the absolute incomes of those currently employed in the F-sector, while it would increase (relatively and absolutely) the incomes of those in the L-sector. Thus those in the F-sector might not agree that the move is towards a more appropriate technology. Only in so far as purely nationalistic objectives were achieved – a reduction in the share of foreigners, and an associated reduction in dependence on advanced countries (see Chapter 5) – might the move be considered appropriate by all groups within a given underdeveloped country; and even in this nationalistic context there are normally some who identify with the foreign interests rather than those of their own country. For these reasons the terms 'alternative' technology, or 'third-world' technology, perhaps, are to be preferred. While the discussion here continues to use appropriate technology, these ambiguities should be borne in mind.

The belief that the technology appropriate to poor countries differs from that appropriate to advanced societies is not, of course, a new one. 'A country which cannot hope to reach within a foreseeable time a capital supply equal per head to that of the United States will not use its limited resources best by imitating American production techniques, but ought to develop production techniques appropriate to a thinner and wider spreading of the available capital.'[2] Various terms have been used to describe the required technology: Schumacher coined the phrase *intermediate technology*,[3] Marsden *progressive technology*, while Mathur discusses 'a third world technology which consists of an adaptation of modern methods to the special conditions of the developing world'.[4] A common term among economists is *labour-intensive technology*. Dickson has promoted the idea of an *alternative technology*.[5]

All these have in common the idea that the third world needs an alternative technology to that of the advanced countries: they differ not only in nomenclature but also in the reasons why they argue an alternative is needed, and in the required characteristics of such an alternative. The rest of this chapter considers the characteristics of what is here described as an *appropriate* (and sometimes *alternative*) technology, drawing occasionally explicitly, but for the most part implicitly, on the pioneering work of the authors just discussed.

The required characteristics of an alternative technology mostly arise in a negative way out of the discussion of the inappropriate characteristics of much technology from advanced countries. Rather than repeat that discussion, Chart 4.1 presents a summary of the characteristics. An important feature of the chart is that it is split into two parts – the left-hand side describes the characteristics in comparison with the type of technology currently in use in the F-sector; the right-hand side presents a comparison with the sort of techniques in use in the rest of the economy. Given the radically different characteristics of techniques in use in the two sectors, whether appropriate technology is described as being more or less capital-using or labour-using, more or less sophisticated or large-scale, etc., depends critically on the comparison being made. There is some tendency to talk exclusively in terms of a comparison with the F-sector – on the assumption, shared by people in many fields, that nothing is happening outside the F-sector. The use of two sets of characteristics is not just a matter of presentation – paying lip-service to the majority outside the F-sector. The nature of the appropriate technology may depend on which sector it is to be associated with, as may the kind of technological research conducted.

Schumacher (and others) have implied that there is a single alternative technology:

> If we define the level of technology in terms of 'equipment cost per workplace', we can call the indigenous technology of the typical developing country – symbolically speaking – a £1-technology, while that of the developed countries would be called a £1000-technology. . . . If effective help is to be brought to those who need it most, a technology is required that would range in some intermediate position between the £1-technology and the £1000-technology. Let us call it – again symbolically – a £100-technology . . . Such an intermediate technology would be immensely more productive than the indigenous technology (which is often in a condition of decay), but it would also be immensely cheaper than the sophisticated, highly capital-intensive technology of modern industry.[6]

Starting with a *tabula rasa* institutionally and technologically one might indeed wish to design and introduce an undifferentiated £100-technology. But one actually starts in a highly differentiated situation, organisationally and technologically. There are therefore in practice at least two approaches to securing more appropriate technology: changing technology in the advanced F-sector, or in the traditional and informal sectors. While the aim would be to bring the two sectors together in terms of investment per head, labour productivity, etc., in the short run the modified techniques would differ according to which sector they are to be associated with. Practical examples of these differences can be seen in a comparison of the work of the Intermediate Technology Development Group, which is primarily concerned with techniques suitable for the rural area, for use on a small scale among the self-employed, and the work of the Strathclyde group on appropriate technology which selects more appropriate techniques for use in the fairly large-scale urban industrial sector.

CHART 4.1. Requirements of a more appropriate technology

| Comparison with | | | |
|---|---|---|---|
| F-sector | | L-sector | |
| *Characteristic* | *Aim* | *Characteristic* | *Aim* |
| Reduced $\frac{I}{L}$ | increase employment opportunities | Raise $\frac{I}{L}$ | increase labour productivity |
| Large or small scale, small preferred | reduce size of entrepreneurial requirements; increase access to local entrepreneurs | Small scale | keep technology within capacity of local entrepreneurs |
| Simplicity of operation and repair | reduce skill requirements | Simplicity of operation, organisation and repair | minimise skill and entrepreneurial requirements |
| Appropriate products | eliminate 'excess' standards | Appropriate products | extend production activities of L-sector to fulfil more needs |
| Urban and rural | | Urban and rural – emphasis on rural | provide opportunities where people are |
| Use of locally produced inputs (raw materials and machinery) | increase local impact, reduce foreign dependence | Use of local inputs | stimulate other local activities |

NOTES ON CHARACTERISTICS IN CHART 4.1

The first requirement is in terms of investment requirements per employee, the requirement discussed by Schumacher above. Because this is the most obvious way in which technology from advanced countries is inappropriate, some have placed exclusive concentration on it. The high levels of investment per man required by advanced-country technology is at the heart of the imbalance of resources, and labour productivity between the two sectors, as discussed in the last chapter; hence a prime requirement of more appropriate technology is that it should permit a more balanced pattern of development, by reducing investment requirements per employee in the F-sector and increasing investment in the rest of the economy. The calculations in the last chapter illustrated the point, showing, on very rough assumptions, the proportion of the workforce that could be employed in the F-sector, given the available savings and the investment per employee of the foreign technology.

The calculations below use the same sort of approach to estimate 'appropriate' levels of investment per man, defined as the level of investment per man which is available on *average* for the economy as a whole. Taking $sY$ (income times the average propensity to save), as the total investible resources available, and $s(1-f)Y$, as the total available to equip the workforce, where $f$ is the fraction of investment devoted to infrastructual investment, then if $k^*$ is investment per worker,

$$s(1-f) \,.\, Y = k^* \,.\, N \qquad \text{(i)}$$

and defining average labour productivity $Y/N = o$, dividing both sides by $N$, and substituting in (i),
then

$$k^* = s(1-f) \,.\, o \qquad \text{(ii)}$$

where $k^*$ is average investment per employee,
$s$ = average propensity to save,
$f$ = fraction of investment devoted to infrastructural investment;
$N$ = the work force,

and $o$ = average productivity of the work force.

This represents the average yearly investment per worker, but investments are lump-sum, occurring sporadically, not in small accretions to the equipment of each worker each year. Assume that equipment has a certain life after which it needs scrapping. Let us say that each investment lasts $L$ years. Then the total resources available, on average, for equipping each worker are $s(1-f) \,.\, o \,.\, L$, or the annual resources available times the number of years the investment lasts. Table 4.1 shows the estimated value of capital equipment per head expressed in terms of average productivity, on alternative assumptions about the values of $s$, $f$ and $L$.

TABLE 4.1

$k^*/o = s(1-f) L.$ on alternative assumptions

| | | $L = 20$ | | $L = 40$ | |
|---|---|---|---|---|---|
| $f$ | $(1-f)$ | $s =$ ·1 | $s =$ ·2 | $s =$ ·1 | $s =$ ·2 |
| 0·2 | 0·8 | 1·6 | 3·2 | 3·2 | 6·4 |
| 0·5 | 0·5 | 1·0 | 2·0 | 2·0 | 4·0 |

The table shows that the average capital/output ratio for a country as a whole, excluding infrastructural investment, varies between 1 and 6·4 depending on the values of $s$, $f$ and $L$.

Table 3.8 showed how $s$ varied for selected countries between 1966 and 1968. 9·8 per cent of developing countries had a gross investment ratio[7] of under 11 per cent; 67 per cent had a gross investment ratio of between 11 and 19·9 per cent and 23 per cent had a ratio of 20 per cent or more.[8]

Expenditure on infrastructure has to be deducted from total savings before arriving at the sum available for investment in equipment and buildings per man. This is *not* because it is assumed that infrastructural investment is unproductive – it is often an essential precondition for many other types of investment – but simply because it is here viewed, definitionally, as a distinct call on resources. Such infrastructural investment includes expenditure on housing, schools, roads etc. The amount spent varies according to the stage of development and the economic and social policy of the country.

Table 4.2 provides approximate estimates of $f$ for three countries.

The classification used in this table follows U.N. statistics, which is

TABLE 4.2 Infrastructural investment as a proportion of total investment

| *Country* | *Residential buildings* | *Other infrastructural building*[a] | *Total* |
|---|---|---|---|
| Mauritius | 20·7 | 22·8 | 43·5 |
| Puerto Rico | 23·5 | 36·8 | 60·3 |
| Canada | 22·0 | 26·1 | 48·1 |

[a] classified in the U.N. statistics as 'other construction'.

SOURCE *Yearbook of National Accounts Statistics, 1972.*

not quite the same as that needed here: the estimates for the three countries probably understate the infrastructural expenditure element in investment, while the figures for these countries may be a poor guide to experience elsewhere.

*L*, the typical life of equipment, depends on a number of factors[9] – physical durability, repair and maintenance requirements, economic obsolescence which is associated with technical and economic change throughout the economy. Estimates of length of life of assets tend to be unreliable.[10] Any actual figures must refer only to assets whose lives have already expired: for current assets estimates are necessarily based on expectations. In developing countries where recent investment accounts for a much larger proportion of the total stock of buildings and equipment, the evidence is even sparser.

The U.K. statisticians assume lives for plant and machinery of between 16 and 50 years, and average life of buildings of 80 years, in estimating capital consumption in manufacturing.[11] Estimates for the United States[12] are for lengths of life between 15 and 22 years for manufacturing equipment, and 40 years for buildings.

It is difficult to be sure whether length of life is likely to be longer or shorter in underdeveloped countries than in developed countries. On the one hand, possibilities and standards of maintenance and repair are likely to be lower. On the other, the forces making for economic (as opposed to physical) obsolescence may be weaker.[13]

As can be seen from this discussion, considerable variation is possible in the variables $s$, $L$ and $f$ determining the aggregate ratio between investment resources per head and output per head. There may be interrelationships between the variables, with, for example, a high $s$ being associated with a high scrapping rate and low $L$. Table 4.1 calculated the ratio $k^*/o$ on alternative assumptions of $s$, $L$ and $f$. In that table the highest value of the ratio was 6·4, the lowest 1·0. Table 4.3 applies these high and low values to actual figures of output per man in a few countries, thus providing high/low estimates of $k^*$ for these countries. We may interpret $k^*$ as appropriate levels of investment per man – being the level of investment expenditure per man that a country could afford for *each* man (on the alternative assumptions of $s$, $L$ and $f$) if investment resources were spread evenly over the whole workforce.

One cannot of course, interpret these figures to mean that each piece of investment should be in line with the appropriate figure. For technical reasons alone some variation is likely. But if some sectors have investment per head far in excess of the appropriate figure, then they will take a disproportionate share of the total investment resources, which occurs when technology designed for the 'appropriate' levels of investment per man of the developed countries is adopted in much poorer countries.

Techniques designed in developed countries are broadly in line with what is appropriate for them: this is indicated by the figures for the

TABLE 4.3 'Appropriate' levels of investment per man, 1970 ($)

| Country | (1) 'high' value of k*/o | (2) 'low' value of k*/o | (3) Average of (1) & (2) |
|---|---|---|---|
| Chile | 18,180 | 2840 | 7670 |
| Argentina | 17,810 | 2780 | 7510 |
| Mexico | 15,870 | 2480 | 6700 |
| Jamaica | 14,020 | 2190 | 5910 |
| Algeria | 9540 | 1490 | 4020 |
| Brazil | 8770 | 1370 | 3700 |
| Zambia | 8640 | 1350 | 3650 |
| Sri Lanka | 3140 | 490 | 1320 |
| Bolivia | 2690 | 420 | 1130 |
| Nigeria | 2620 | 410 | 1110 |
| Thailand | 2370 | 370 | 1000 |
| Botswana | 1980 | 310 | 840 |
| India | 1860 | 290 | 780 |

| | 'high' value of k*/o | 'low' value of k*/o | Average of (1) & (2) | Actual k/L (excluding working capital) |
|---|---|---|---|---|
| U.S.A. | 75,070 | 11,730 | 31,670 | 27,104[b] |
| Australia | 44,420 | 6940 | 25,700 | |
| France | 43,520 | 6800 | 18,360 | 14,686[b] |
| W. Germany | 40,960 | 6400 | 17,280 | |
| U.K. | 29,630 | 4630 | 12,500 | 11,940[a] |
| Japan | 24,190 | 3780 | 10,210 | |

[a] Fixed capital per man in manufacturing, 1974 at 1970 prices, using £1 = $2·4 exchange rate.

[b] Based on U.K. figures applying ratios for net stock of non-residential structures and equipment, in Denison (1967), p. 166.

SOURCE Tables 3.8, 4.1 and 3.7 above.

U.K., U.S. and France in Table 4.3, which shows actual fixed capital per man in manufacturing which is roughly equivalent to the 'appropriate' level shown in column (3). However, the conceptual difficulties arising and the dubious sources used in arriving at these 'actual' figures must be taken into consideration.

*Scale*: it is clear from the discussion in the last chapter that the large scale of many modern techniques causes problems, when used in LDCs.[14] This is partly because the size of the market is, typically, much smaller, and partly because of the managerial and financial requirements large-scale production imposes, so that it becomes out of the reach of most potential entrepreneurs. Hence small scale is often emphasised as a leading characteristic of appropriate technology.[15] As far as use in and by the informal and traditional sector is concerned scale is a critical factor. Techniques must be inexpensive absolutely – i.e. per machine – as well as per worker, and must not require large numbers of workers, or large markets, for their success.

The question of appropriate scale for the F-sector is a more difficult one. In this sector, relatively large-scale units of production already exist; modifications to the technology used in a more appropriate direction, for example, to make greater use of labour and of local materials, might be beneficial, even without any modification of size. Those who have concentrated on the need for more labour-using technology have had this type of modification in mind, and have regarded scale, in itself, as irrelevant.[16] However, *ceteris paribus*, small scale is more appropriate than large in most countries, given the size of market and local systems of organisation.

*Rural/urban*: the rural/urban question is, to some extent, tied up with that of scale, as really large-scale production can only take place in an urban setting, Those who have emphasised the need for small-scale techniques have also tended to emphasise the need for additional techniques in the rural areas:

> As we see it, the greatest task is to redress the balance between rural life and city life, by going into the rural areas with appropriate technologies of self-help so as to foster hope and self-reliance. If the breakdown of rural life continues there is no way out: . . . I have often said to friends of mine in developing countries, countries which are called thus but stubbornly refuse to develop, that the cause of their country's misery is not backwardness but *decay*, *the decay* of the rural structure.[17]
>
> . . . The task then is to bring into existence millions of new workplaces in the rural areas and small towns . . . workplaces have to be created in the areas where the people are living now, and not primarily in metropolitan areas into which they tend to migrate. (Schumacher, 1974)

The majority of people in underdeveloped countries live in the rural areas, as shown below – so, potentially, techniques designed for rural use might extend to more people than urban techniques. Moreover, despite considerable urban squalor and poverty, poverty is generally concentrated in the rural areas in greater proportion than population. The manifold problems arising from rapid urbanisation also suggest that more attention needs to be paid to creating rural opportunities. All the theories of rural/urban migration agree that rural/urban opportunities (or lack of them) play a significant role in influencing the movement to the towns. None the less, the rural aspects of appropriate

TABLE 4.4 Urban/rural population, 1970 (millions)

| | *Urban* | *Rural* |
|---|---|---|
| Africa | 75 | 277 |
| Latin America | 161 | 123 |
| China | 167 | 605 |
| South Asia | 231 | 880 |

SOURCE *Development Cooperation, 1975 Review* (OECD, November 1975) Table IV–I.

technology may have been overemphasised. Urban poverty and lack of productive opportunities is also widespread. While the balance of numbers suggests that comprehensive improvements in rural techniques would affect more people, the problems of developing such techniques and communicating them throughout the rural areas might well be greater. In so far as modification of F-sector technology is concerned, this largely has to be urban because that is where the existing techniques are. External economies are often greater in an urban setting.

The choice between urban and rural techniques, in so far as there is a choice, is not therefore simply a matter of absolute numbers, or absolute numbers in poverty in the two sectors, but also how many the new techniques are likely to reach, and their income-creating implications.

*Simplicity*: intermediate machines are *simple*.[18] The concept is not a simple one. In this context it can be interpreted in terms of educational and training requirements, the simpler the machine the less the demand for skills requiring education and training which are scarce in most developing countries. There are a number of ways in which a machine can be simple:

1. simple to make;
2. simple to operate;
3. simple to repair and maintain;
4. organisationally simple.

None of these four types of simplicity are necessarily associated with each other – indeed some of them are likely to involve complexity in others.

The relationship between different characteristics is in part a technical matter, but also, as argued earlier, a matter of the historical development of the techniques. Historically, skills of one type – those of craftsmen – have tended to be displaced by machinery and have disappeared, while other skills required for the operation of the new machines have replaced them. The oldest craft-type techniques tend to be labour-intensive, but also craft-skill-intensive. Techniques originating from the early industrial revolution period require unskilled factory labour and some mechanical skills. More recently developed techniques require considerable skills for their management, upkeep and operation – very different types of skill from the early craft techniques.

Generally speaking the greater the division of labour the greater the simplicity of operation (as Adam Smith pointed out). In terms of operational simplicity, therefore, there may be a conflict between the desire for simplicity and the desire for small-scale production. Moreover, administrative complexity may be increased as the division of labour, and simplicity of operation, increases. Organisational or administrative complexity may be internal or external. Internal complexity is likely to increase with the division of labour within the firm – since coordination between the different operations is required. External complexity occurs with increasing division of labour between firms and requires the existence of a network of firms that are efficiently coordinated by the market (or some substitute body). Complexity of trading operations replaces complexity of coordinating internally. A small-scale specialist firm may operate within a well-developed market with few resources devoted to administration and coordination. However, in the absence of such a market the small firm may be less well placed than the large firm since it cannot rely on the market for the necessary coordination, while the large firm operates on a do-it-yourself basis as far as coordination is concerned.

Maintenance simplicity is of particular importance in a developing country where specialist mechanics may be few and far between. Here the relativity of the concept is relevant. In some cultures care and maintenance of machines is inbred,[19] but other forms of maintenance, e.g. of animals, requires considerable training. Elsewhere the opposite situation may hold. Production using appropriate technology requires that huge numbers of separate units somehow get off the ground. This is unlikely to happen naturally and without considerable organisation and administration in terms of spreading knowledge about the techniques, providing access to capital, to inputs and to markets. The administration required per unit of output is almost certainly greater than for Western technology. This is partly because of the far greater spread, both geographically and in terms of numbers, of those who have to be contacted and organised. It also arises because there is greater experience and knowledge of how to organise production on Western technology lines. People qualified and experienced in the managerial problems of Western technology are more widely available in the developed countries and in the developing countries, while little experience in organising appropriate technology exists.

*Appropriate products*: an essential characteristic of appropriate technology is that it produces products appropriate to the levels of living of the economy in which it is to be used. As far as consumer goods are concerned, this means simple products designed for low incomes, which fit in, in a balanced way, with the life-styles of the people for whom they are intended. Most products are inputs into further productive processes: these intermediate products are appropriate if they may be used as inputs into appropriate productive processes.

*Local inputs and production of machines*: this requirement is intended to maximise the linkage effects of the technology on the rest of the economy. Local production of the machinery can be of particular significance because of the important role of local capital goods industries in promoting innovation and skills.[20]

The characteristics summarised in the chart, and discussed above, represent the direction in which technology needs to go, if it is to be appropriate for underdeveloped economies, to enhance incomes in these economies by making use of all their resources, instead of distorting patterns of development in the way that techniques from advanced countries do. It must be re-emphasised that many techniques may improve on advanced technology in some respects but not in others, so that in particular cases it may not be possible to classify techniques as 'appropriate' or 'inappropriate'. There have been many disputes as to whether particular characteristics are or are not necessary for appropriate technology – e.g. are large-scale labour-intensive techniques producing inappropriate products for export appropriate? The answer is that they are in some respects, but not others, and any decision about whether to use them or not must depend on the alternative available.

Any comparison of actual techniques will quickly bring considerations of efficiency into discussion. So far efficiency has not been mentioned as a required characteristic. There are problems of definition here which need discussion. It is often claimed that alternative techniques are 'inferior and outdated'.[21]

> You should not go deliberately out of your way to reduce productivity in order to reduce the amount of capital per worker. This seems to me nonsense because you may find that by increasing capital per worker tenfold you increase the output per worker twentyfold. There is no question from every point of view of the superiority of the latest and more capitalistic technologies. (Kaldor (1965))

The point being made is that alternative more labour-intensive technologies, while reducing investment per man, increase investment per unit of output. Hence their selection involves an immediate loss of output, and subsequently, because of the lower level of income and therefore savings, of investment and employment too.[22] The historical development of techniques in advanced countries has tended, as argued in Chapter 1, to bring this situation about – not because more-investment-using techniques are in their nature more productive than more-labour-using techniques, but because *historically* the more-investment-using techniques were developed later, and therefore embody more knowledge than the earlier more-labour-using techniques. But this does not invariably mean that the earlier techniques involve lower levels of output than the later techniques – as many empirical examples have shown.[23]

It certainly does not mean that labour-using techniques, if developed today using today's knowledge, need involve sacrificing output. However, it does suggest that techniques possessing all the appropriate characteristics listed above may still not be appropriate, or the right ones to use in practice, if they are shown to be inferior in this way. Hence the relative efficiency of the technique must also be considered. Here, however, we come up against the definitional problem. What exactly is meant by relative efficiency?

While the notion of inefficiency, or inferiority, contained in the quotation above, that of lower output with the same investment, sounds straightforward, on closer examination it is not. In the first place the nature of the output normally differs between techniques. Indeed, by requiring appropriate techniques to produce appropriate products, as compared with inappropriate techniques, this difference has been built into the comparison. In the second place, the output from the techniques and the income arising from the productive process are differently distributed. Inappropriate techniques are associated, as argued in Chapter 3, with unequal distribution of income and imbalanced consumption patterns: much of the benefit from the techniques goes to foreigners in the form of technology payments, managerial payments, payment for imported inputs, and sometimes for dividends too. As far as the local population is concerned the benefit is mainly confined to skilled workers and relatively high-wage unskilled workers – the 'aristocracy of labour'.[24] In contrast, appropriate techniques are used by and create income for the poorer people in the country. Leakages to foreigners and to the local elite, ideally, are small. Thus even if in some technical sense there was a loss of efficiency – let us say less yards of cloth per rupee's investment – it does not follow that the more efficient technique should be adopted. This is true in the long run too, because the reinvestment that highly productive techniques may make possible (if sufficient of the profits are retained in the country for reinvestment) may be reinvestment in further inappropriate techniques, which, while possibly resulting in a high statistical growth rate, may still be confined in its income effects to those with above-average incomes.

This discussion does not mean that considerations of efficiency should be totally ignored. *Ceteris paribus*, the more efficient technique is to be preferred. But *ceteris non paribus*, the differences in circumstances, and particularly differences in income distribution effects, must also be considered.

One proposal often made is that underdeveloped countries, lacking appropriate techniques developed for them, should turn to older techniques from developed countries.[25] Such techniques, as argued in Chapter 1, should be more appropriate to poor countries because they are more labour using, and were designed to produce products for poorer consumers than techniques developed in advanced countries today. But there are difficulties about the use of older techniques. Most obviously they are no longer in use or production, so their use would require redevelopment. It might be better to devote resources to the development of new techniques. Secondly, the older techniques may require materials and skills that are no longer available – the technology package the techniques were designed for has largely disappeared. Thirdly, having been developed earlier, they tend to be relatively inefficient as compared with new techniques, since they come at an earlier stage of scientific and technical know-how. Fourthly, while the early stage of the industrialised nations was in some respects similar to that of poor countries today, in many respects it was dissimilar. Most obviously, the climatic differences of today were also present a hundred years ago. There are also economic differences. The rate of growth of population in most underdeveloped countries today is at least twice that of nineteenth-century Europe. The ratio of people to land, and people to savings, is also higher. Cultural and institutional differences are great.

### IS APPROPRIATE TECHNOLOGY POSSIBLE?

The introduction of a more appropriate technology would reduce the large disparities in control over resources, labour productivity and income distribution between different parts of the economy. The dualism produced by advanced country technology would ideally, give way to a unified and balanced form of development. The major problems resulting from the dualistic pattern – particularly those of imbalance in employment opportunities, the appearance of open unemployment as a chronic problem, and maldistribution of income – should disappear if appropriate technology, equally accessible to all, were introduced. If the technology were efficient, the absolute level of incomes should also rise as more of the community's previously unused resources – of labour and of natural resources – are brought into use.

All this sounds too good to be true. There are major difficulties which prevent this being the immediate solution to all development problems.

First, the historical (and current) development of technology has favoured technology suited to the advanced countries, as discussed in Chapter 1. This is not just a question of resources going to the development of techniques in and for advanced countries, but even the R and D carried out in third world countries is mainly devoted to the development of the sort of techniques in use in the 'modern' sector those countries, which means similar techniques to those developed in th advanced countries.[26] Thus the only 'appropriate' techniques available today are techniques developed in the traditional sectors of poor countries, old techniques from advanced countries, and a few newly developed, produced by those currently searching for appropriate technology. While each of these three do provide genuine technical alternatives in some cases, their number is limited, and often, because of the few resources devoted to their development, they are relatively inefficient. Chapters 8–10 present some empirical evidence of the range of techniques and their characteristics in some industries. The conclusions from these chapters are two-fold: one, that there is much more choice available than many have assumed. It has been established that in a large number of industries there are more appropriate alternative techniques which can, in certain economic conditions, compete successfully with the inappropriate techniques. There are a growing number of recently developed appropriate techniques, designed to be operated on a small scale, in a labour-intensive manner and sometimes producing more appropriate products. However, empirical studies are still limited in their coverage, and the industries which have not been studied may well prove to offer less alternatives than those that have.

The second conclusion from the studies is that, despite the existence of more appropriate alternatives, inappropriate techniques continue to be used in many parts of many poor countries. The reason for this is partly that the economic and social structure of the countries favours the choice of inappropriate techniques; however, in some cases the relatively low efficiency of alternative techniques is responsible. Appropriate techniques have inevitably suffered, as to their existence, their number and their productivity, from their historical neglect in the development of technology. Moreover, administrative difficulties associated with the widespread introduction of such alternative techniques may further reduce their relative efficiency. Hence a major scientific and technical effort needs to be made in developing appropriate techniques across the board and improving their productivity.[27]

Too much can be made of the so-called 'existence' question.[28] This is because the question of whether or not an 'efficient' alternative technology exists is closely tied up with the whole strategy of development, and can only be assessed within the context of a particular strategy. For example, alternative techniques do exist and/or could be readily developed if a rural-oriented, small-scale, self-reliant pattern of development is adopted, but if the strategy consists in rapid industrialisation which will permit the production of goods that can compete internationally with goods produced by advanced countries, and will serve domestic high-income consumers in competition with products from advanced technology, then advanced-country technology is required.

The social, economic, political and technological system in being in large part determines the choice of techniques: in a society which has already adopted inappropriate technology, alternative techniques tend to appear inefficient, and, even where apparently efficient, they are often rejected in favour of the further use of inappropriate techniques (as shown in the empirical evidence) because of the links between different parts of the system.

One such link arises from the package aspect of technology. Thus, if in general advanced techniques are used, this creates a major bias in favour of the selection of other advanced techniques. Alternative technology therefore needs to be introduced as a system, if it is to be efficient. But this does mean – for most development strategies – that the existence problem becomes critical. For while undoubtedly bits and pieces of an alternative system have been developed, it is premature to claim that an alternative system as a whole already exists. It is probably for this reason that the Chinese have walked on two legs, introducing appropriate technology to one part of the system, separated from the rest of the economy. The kind of separation required – of markets, sources of supply, etc. – is much more difficult to achieve in market economies: hence the destruction of traditional appropriate activities by the introduction of advanced-country factory methods, as for example in the shoe industry in many countries, or in maize milling as shown in Chapter 9.

Further links arise from the relationship between the technological system in use and the selection mechanisms, discussed in previous chapters. Selection mechanisms tend to favour the type of technology in use. F-sector technology tends to produce a type of decision maker, informational structure, market structure, set of relative prices and access to resources, all of which favour the use of F-sector technology and make an alternative technology appear economically inefficient. Once a particular system is established it thus generates its justification, making it increasingly difficult to establish an alternative system: such an alternative would be justified with a different set of decision makers, with different objectives and different economic conditions – but the alternative type of system would only emerge once the alternative technology system is in use.

Implicit behind some of this discussion is what might be termed the political economy of technical choice. The political economy of a system may be defined as the distribution of the control over resources – both consumption and investment – to which it gives rise. Associated with each technique is a particular distribution of benefits, for the producers associated with the technique and for the consumers of the product. While some variation is possible in theory, so that a particular investment may be taxed or controlled to offer a different distribution of benefits, in practice there is a strong connection between the technique chosen and the distributional consequences. The connection is stronger between a technological system as a whole and its political economy. Advanced-country technology concentrates its benefits in the advanced countries themselves – which supply the technology and, to a varying degree, the machinery and the management, and which often own the equity and receive the profits; it also generates income among the minority in the poor countries who work for, sometimes manage, and consume the products of the technology. The distribution of benefits from appropriate technology is entirely different. The beneficiaries are small-scale local producers, the potentially un- or underemployed, and low-income consumers who provide a market for the low-income appropriate products. Appropriate technology is designed (and only qualifies as appropriate if its succeeds) to benefit the majority who are largely left out of the process of development with inappropriate technology. There is thus a stark conflict in terms of political economy. The gainers from advanced-country technology – advanced-country suppliers and the elite within the underdeveloped country – are the losers from appropriate technology. But control over resources is determined by the system of technology in use. It is this which provides the surplus for investment, and the markets for consumption goods, determining which types of technique are most profitable. The political economy of technical choice is such that each system tends to be self-perpetuating. There is a process of cumulative causation. The use of advanced-country technology builds up biases in selection mechanisms in favour of its further use; this in turn leads to further decisions in favour of such technology. Moreover, the continued use of such technology increases its relative efficiency as information, learning by doing and scientific and technical innovation are related to the system in use, while appropriate techniques are neglected.

The initial historical bias in technological development towards technology designed for advanced countries thus sets up forces, in poor countries as well as rich, which tend to perpetuate the use of such techniques and discourage the adoption of an alternative system of technology. To develop an alternative system requires control over scientific and technical resources which, for the most part, the underdeveloped countries lack; it also requires a redirection of income and investment resources away from those currently benefiting from advanced-country technology, both in rich countries and in poor. Such a redirection is itself very difficult to achieve unless an alternative technology has already been introduced, because the distribution of the income is a consequence of the system in use.

There are two alternative cumulative development cycles: the inappropriate technology cycles (*A* in Figure 4.1), and the appropriate technology cycle (*B*). Advocates of appropriate, or intermediate, technology see the need to make the jump from *A* to *B*. They do not explain

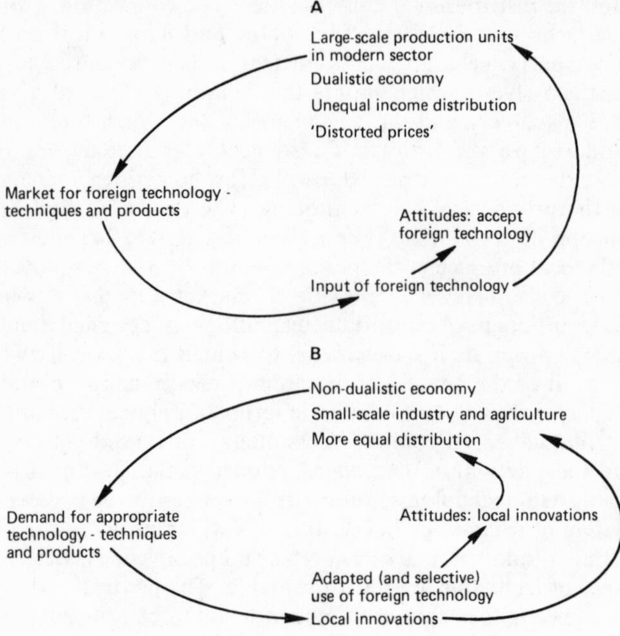


FIG. 4.1

how such a jump can or will be made, given the existence of cycle *A*, reinforced by advanced-country interests in maintaining it. Ultimately the answer is to be found in the realm of politics rather than economics and technology. But the close connection between these three suggest that there are economic and technical preconditions of a successful move – necessary if not sufficient conditions. The next three chapters consider aspects of these preconditions.

NOTES

1. This point is developed in Stewart (1975).
2. Hayek and Lekachman (1955), p. 89.
3. He first used the term while a consultant to the government of India in 1963.
4. Mathur in a speech to the sixth Asian regional conference of the I.L.O., Tokyo, 2–13 September 1968. Extracts appear in the *Bulletin of the Intermediate Technology Development Group*, No. 6 (October 1969).
5. Dickson is concerned primarily with an alternative for developed countries; Schumacher (1973) shares this concern with finding alternatives for developed countries as well as underdeveloped.
6. Schumacher (1973), p. 150. Schumacher calls his figures 'symbolic': actually they are wrong, being far too low for both modern and traditional sectors, but the point remains.
7. Ratio of gross domestic fixed capital formation to gross domestic product.
8. See United Nations, *World Economic Survey*, 1969–70, Table 29.

9. The empirical case studies in this book (Chapters 9 and 10) discuss length of life of assets in particular cases.
10. For example, much of the difference between Redfern's (1955) and Barna's (1957) estimates of capital stock in the U.K. were due to differing length of life assumptions – see Dean (1964).
11. See H.M.S.O. (1968), p. 386.
12. B. Creamer, Dobrovolsky and Borenstein, quoted in Stigler (1963), p. 121.
13. Strassman (1968) states (p. 201): 'Almost all Mexican sample enterprises kept equipment going longer than would have been true in a high-wage, capital-abundant economy.' However, Lewis (1955) argues that depreciation rates are much *higher* in less developed countries because of a higher rate of physical deterioration (p. 203–4).
14. This is the inappropriate characteristic of advanced-country technology most emphasised by Merhav (1969).
15. See the publications, for example, of the I.T.D.G.
16. Broadly, this appear to be true of the Strathclyde study and some other writings, e.g. Ranis (1973) and Pack (1974). The assumption of divisibility (behind much neo-classical thinking) excludes scale as a consideration, since any technique can be used at any scale of production.
17. Schumacher in *Enterprise* (1969).
18. Schumacher (1974): 'The production methods must be relatively simple, so that demands for high skills are minimised not only in the production process itself but also in matters of organisation, raw material supply, financing, marketing and so forth.'
19. This was emphasised, for example, by an advertisement for TWA which showed a picture of a small American boy with the caption, 'You happen to be looking at a very good mechanic ... In America kids play baseball. In America kids learn about cars and engines.' (*Sunday Times*, December 1969).
20. See the discussion in Chapter 6.
21. Quoted by Schumacher (1974), p. 151.
22. The conflicts between output and employment objectives such a situation would give rise to are discussed in Stewart and Streeten (1972).
23. See Chapters 8 to 10, Bhalla (ed.) (1975), and Jenkins (1975).
24. See Arrighi in Arrighi and Saul (1973).
25. E.g. by Pack and Todaro (1969).
26. See the discussion of science policy in Cooper (ed.) (1972) and Cooper (1974).
27. See Bhalla and Stewart (1976) for some proposals for international institutional change to bring about appropriate innovation.
28. Streeten (1973) describes this existence problem as the 'suitability gap' in technology.

# 5 Technological Dependence

The *dependency* theorists[1] are concerned with the whole relationship between advanced countries and third world countries: the dependent relationship is exhibited in cultural as well as economic features of third world countries. From this point of view the *negritude* movement in French Africa was as much a struggle against dependence as the (rather more prosaic) bargaining strategy of the Andean Pact countries. The dependent relationship pervades political institutions and political decision making as well. As a result many countries are incapable of following an alternative path, not only because the world economic facts of life make it impossible, but because the cultural, psychological and economic pressures of the dependent relationship have conditioned decision makers in third world countries so that they do not *wish* to follow an alternative strategy. Many conflicts which appear to be conflicts of interest between advanced countries and under-developed countries become internalised within third world countries, with powerful sections of the community representing the advanced-country interests within third world countries.

The dependency school of thought encompasses many variations.[2] There are those whose thought originates from Marxist analysis. For them, dependency is the inevitable outcome of capitalist development in the advanced countries, and the internalisation of conflicts within the third world represents the Marxist class struggle transferred to the third world. For others, dependency-type analysis has its origin in theories of underdevelopment.[3] The underdevelopment theorists challenged the conventional wisdom of Western economists of the 1950s, as epitomised (at its crudest) by Walt Rostow's stages of growth. This conventional wisdom held that growth and development were unilinear; that there were various stages of development that societies went through on the way to industrialisation. The poor countries of the world were simply a replica of the industrialised countries at a previous stage of development. Given the right sort of conditions they would inevitably follow the same path – and indeed they could be given a boost or leg up through additional resources and know-how available to them from the advanced countries. They had, in Gerschenkron's phrase, the advantages of late-comers.

Built into this approach was the almost Darwinian view, that progress was inevitable – not just in the sense that development would inevitably occur, given the right conditions – but also in the sense that such development would lead to society being better off. Theories of underdevelopment challenged the unilinear and progress view of development. They argued that the third world differed from the advanced countries at an earlier stage *because of the existence of the advanced countries*, and the impact of their society on them. The underdeveloped world economy could not be analysed in isolation from world developments because their economy was in large part conditioned by events in the advanced countries: through trade, migration, capital and technology flows the advanced countries determined the nature of the economy of third world countries. The underdeveloped state of third world countries was attributed not to the fact that they were at an earlier stage of history than the advanced countries, but to the fact that the impact of the advanced countries on the third world had caused their underdevelopment. The nature of the impact of the advanced countries, particularly of the capitalist countries, on the third world, and the terms of such impact, were such as to impoverish the third world economically, culturally and psychologically. Thus it was argued (e.g. by Frank and Griffin) that many poor countries were not only relatively much better off say in the eighteenth century, but *absolutely* per capita income was substantially higher than it is today.

The dependency theorists on the whole accept this view, but add to it the importance of the dependent relationship in determining both what happens and what is possible. Put at its simplest, the dependent relationship makes it impossible for third world countries to pursue policies which would not impoverish them. Their ties are such that effective independent policies are not possible. The first prerequisite of following non-impoverishing policies is then to break the dependent relationship. But even this may not be possible because of the all-embracing nature of dependency preventing any such breakout. Here, perhaps, it is fair to conclude that the Marxist and the structuralist dependency school part company. Some, at least, of the structuralists appear to believe a break is possible, while most of the Marxists would not, considering the relationship as the inevitable outcome of capitalist development – only a change at the centre makes change at the periphery possible.

The dependent relationship means that events in third world countries are determined by what happens elsewhere, notably in the centre. But it also means that parts of the third world – those in what we have described as the modern foreign-technology sector – are more accurately viewed as part of the centre, rather than the periphery, and their actions therefore are taken in the light of the interests of the centre, rather than that of the rest of their own country.

This chapter is concerned with *technological dependence*, its nature and consequences. As we shall see, technological dependence is closely linked to the more general relationship, briefly discussed above. It is linked as a cause, a symptom and a consequence of the general relationship. Indeed, it is possible to argue that technological dependence is the most critical aspect of the whole relationship – so long as it continues it is impossible to break out of the general relationship, and if it could be avoided then genuine general independence would be possible. However, we start by taking a pragmatic look at the concept of technological dependence, treating it independently of the general relationship – a situation that arises from the world imbalance in technological development.

## CHARACTERISTICS OF TECHNOLOGICAL DEPENDENCE

Technological dependence describes a relationship between countries and one which, like the relationship between persons from which it is derived, is not susceptible to rigorous definition. The nature of the concept is that it is imprecise, describing a syndrome of symptoms: this does not mean that the concept is useless – but it does mean that a search for a watertight definition is misplaced.

Technological dependence arises where the *major source* of a country's technology comes from abroad. In the case of third world countries, the major source is advanced countries. The dependence is greater, the greater the extent of reliance on foreign technology, and the more *concentrated* the source. That is to say, a country should be described as more technologically dependent if all its foreign technology comes from a single country, than if its sources are spread among a number of countries. In some cases sources are widely and evenly spread over the economy as a whole, but in each individual industry, the source is concentrated. This too is an aspect of technological dependence.

All societies from subsistence to the most advanced operate a technology, be it implicit or explicit. In some societies there are more formal means for transmitting this knowledge than in others. In some it is largely a question of word of mouth and imitation; in others complex systems of education have developed, and technology has been *commercialised*.[4] Commercialisation of technology occurs when it becomes part of a system of property, and its transfer is no longer free; the knowledge is monopolised and bought and sold. The commercialisation of technology developed along with the rapid increase in technological change, in the industrialising countries in the nineteenth and twentieth centuries.

In most societies, a whole variety of means of acquiring and transmitting knowledge coexist. Informal and traditional means of transferring technology continue for some areas of life, and in some sectors of the economy, while in the 'modern' parts of the economy commercialisation of technology has taken root.[5] This coexistence of different types of transmission is true even in the most advanced economies, where commercialisation has penetrated most widely and deeply. In developing countries, where in many cases the 'modern' sector covers only a minority of total activity, informal systems of transmission form a correspondingly large part of total technology transmission.

The formal sector covers those activities organised, broadly, on the lines of activities in advanced countries, and using techniques and producing products identical (or very similar) to those of advanced countries. For the most part, the technology in use in this sector is transferred, in one way or another, from advanced countries, though some may be locally produced, along similar lines, as a result of local research and development. Technological dependence describes this situation of almost exclusive reliance on advanced-country technology, with a little adaptation, *in* the formal sector. The informal sector covers activities in the rest of the economy which have not been absorbed wholly into the formal sector, or, put in another way, into the advanced countries' industrial system. In this sector the sources of technology are diverse, as are the methods of transfer. As before (Chapter 2) one can usefully distinguish between the traditional-informal sector and the modern-informal sector. The former includes all those employment activities, agricultural and non-agricultural, which developed in the traditional society before the penetration of advanced-country, post-industrial revolution methods. By definition, the source of technology for this part of the sector is local tradition, and the method of transmission of such technology is also along traditional lines. In contrast, the modern-informal sector represents new activities that have developed since, and in response to, developments of the formal sector. The technology in use in this part of the economy is an amalgam of traditional methods, locally generated methods and technology derived from the formal sector, but often adapted. The methods of transfer of technology in this sector are invariably informal; there are no explicit technology contracts, nor payments for technology. Within the formal sector too. there is a wide variety of means of transferring technology, within and between countries.[6]

A good deal of the technology in the formal sector is derived from informal contacts. For example, tastes and customs may be transferred through contact between peoples, where neither payment nor contracts are involved. Much know-how, as opposed to know-that, is transferred through learning on the job, though some is transferred via formal training programmes. Technical information is supplied through 'exchanges of books, learned journals, trade journals, and sales literature'.[7] The sale of machinery provides a major source of transfer. The machinery normally comes with instructions on installation, use and repair. Foreign investment forms another major source of a wide variety of technology transfer, including proprietary process knowledge, product specifications, trademarks and brand names, management systems, training, learning on the job, etc. Some of this transfer is covered by formal patents, licences and payments for R and D and management services; the rest comes as part of the package.

We argued earlier that technological dependence occurs where a major source of technology comes from overseas. Most advanced nations specialise and trade in technology, so many advanced nations could, in this sense, be described as being technologically dependent, importing most of their technology and exporting very little, e.g. Canada and New Zealand. Many other countries export and import technology. Technological dependence (as opposed to interdependence) is partly a matter of the balance of this trade; if the vast majority of technology is imported, and little exported, then the country is dependent. On the other hand, if the trade is fairly evenly balanced, this is not so. Some countries which are heavily net importers on the technological balance of payments, e.g. Japan, may none the less be clearly a good deal less dependent on foreign technology than other countries in a similar *statistical* position. This is because the dynamics of the situation are relevant as well as the statics: i.e. the way the balance is changing over time. Thus, although Japan is in heavy deficit her exports of technology are increasing much faster than her imports.[8]

Countries which are heavy importers and exporters of technology, like most advanced countries, are clearly in some sense dependent on foreign technology just as they are on foreign goods, yet there is a big difference between this situation and one where countries are in heavy net deficit for technology. In the former case – the two-way relationship between technology flows – countries may gain from technological specialisation that trade makes possible, without suffering some of the evil consequences of excessive technological dependence. Since they export and import technology they are in a much stronger bargaining position in determining the terms on which they import the technology. In any case countries that export and import technology may gain on the roundabouts (exports) what they lose on the swings (imports) in payments made for technology. In contrast, third world countries which are heavy net importers of technology – one way relationship countries – are much more likely to get their technology on poor terms.[9]

Another dimension of dependence is the extent of flexibility, and substitutability of local for foreign resources. Specialisation and trade in technology, like goods, may make for a more efficient allocation of technological resources and increase the returns to these resources. Where specialisation leads to rigidity and incapacity to substitute local resources for foreign, should this become necessary, the specialisation is responsible for a form of dependence. In general, it is widely believed (though rarely empirically shown) that in the production of goods developed economies exhibit far more flexibility than underdeveloped economies. In the production and use of technology, this seems even more likely to be the case, since the technological resources available in most underdeveloped countries are so small. Again this distinguishes the two-way trade countries from the one-way trade countries. Countries which export technology as well as importing it are more likely to be able to switch their technological resources to import substitution should this be necessary. This too strengthens their bargaining position in determining the terms of technology imports.

It is possible to overstate the inflexibility and incapacity of local technological resources in underdeveloped countries. Foreign technology imports have an inhibiting effect on local technology development, as will be discussed later. Technology cut-off can have an unexpectedly stimulating effect on local capacity, as the reduction of Russian technical aid and assistance to China in the 1950s showed.[10]

There are other dimensions of technological dependence, besides the over-the-counter imports and exports. Thus Cooper and Sercovich described the third world as being in double dependence because 'the elements of technical knowledge themselves have to be transferred, but so does *the capacity to use this knowledge in investment and production*.'[11] This second element of dependence, clearly absent in Japan and in most advanced countries in deficit on the technological balance of payments, but present in many developing countries, is responsible for a major difference in the extent of their technological dependence, irrespective of what the figures of technological payments show.

The significance of the figures for imports of technology also depends on what the country does with the technology imported. Here the contrast between India and Japan is instructive. Whereas in India foreign technology imports often provide a substitute for local technology resources, and rarely an impetus to it,[12] in Japan imports of technology have been strictly controlled so as to avoid competing with local technology, and those technology imports that are allowed have been used as the basis of further local innovation. Thus research and development expenditure on imported technology forms one third of Japan's total R and D.[13]

### SOME INDICATORS OF TECHNOLOGICAL DEPENDENCE

Technological dependence arises initially from the imbalance in technological *capacity*, i.e. the capacity to produce technology. Different indicators of this imbalance are contained in Table 5.1.

The imbalance is also indicated by figures for patents issued: only 6 per cent of the estimated $3\frac{1}{2}$ million patents in existence in 1972 were granted by developing countries, and less than one-sixth of that total were owned by nationals of developing countries.[14]

The extent of reliance on imported capital equipment is an indicator

TABLE 5.1 Technological capacity: selected indicators
(Averages expressed as medians for 1970 or latest year available)

| | *Developed market economy countries*[a] | *Developing countries and territories* | | |
|---|---|---|---|---|
| | | *Africa*[b] | *Asia*[c] | *Latin America*[d] |
| I SCIENCE AND TECHNOLOGY | | | | |
| (i) Ratio of total stock of scientists and engineers per 10,000 pop. | 112 | 5·8 | 22·0 | 69 |
| (ii) Ratio of technicians per 10,000 pop. | 142·3 | 8·3 | 23·4 | 72·2 |
| (iii) Scientists and engineers engaged in R & D per 10,000 pop. | 10·4 | 0·35 | 1·6 | 1·15 |
| (iv) Technicians engaged in R & D per 10,000 pop. | 8·2 | 0·4 | 0·6 | 1·4 |
| (v) Expenditure on R & D as percentage of GNP | 1·2 | 0·6 | 0·3 | 0·2 |
| II HIGH LEVEL MANPOWER | | | | |
| (vi) Professionals and technicians as percentage of economically active population | 11·1 | — | 2·7 | 5·7 |
| (vii) Percentage of the economically active population employed in manufacturing sector | 25·4 | 3·5 | 10·5 | 14·1 |
| (viii) Literacy rates (per cent) | 96[e] | High[e] 20 Low[f] 15 | 32 | 77 |
| (ix) Ratio of primary and secondary enrolment to school age population | 92[e] | 32 | 56 | 78 |

[a] The size of the sample in this column varies by indicator, ranging from four countries in line (ii) to 25 countries in line (ix).

[b] The size of the sample in this column varies by indicator, ranging from eight countries in lines (i) and (ii) to 46 countries in lines (viii) and (ix).

[c] Excludes China. The size of the sample in this column varies by indicator, ranging from seven countries in line (vi) to 36 countries in lines (viii) and (ix).

[d] The size of the sample in this column varies by indicator, ranging from seven countries in lines (i) and (ii) to 43 countries in line (viii).

[e] Includes Greece and Turkey.

[f] Taking upper limit of estimates where no precise figures were given, e.g. for 10–15 per cent, 15 per cent would be used for high estimate and 10 per cent for low estimate.

SOURCE *Transfer of Technology, Technological Dependence: its Nature, Consequences and Policy Implications*, Report by the UNCTAD Secretariat, TD/90, December 1975.

TABLE 5.2 Share of imports of capital goods* in gross fixed investments

| DEVELOPING COUNTRIES | *Early 1960s* | *Early 1970s* |
|---|---|---|
| Korea | 19 | 33 |
| Iran | 36 | 49 |
| Nigeria | 47 | 44 |
| Thailand | 43 | 37 |
| Malaysia | 62 | 63 |
| Panama | 54 | 41 |
| Tanzania | 49 | 53 |
| Mexico | 27 | 24 |
| Brazil | 20 | 17 |
| El Salvador | 41 | 47 |
| Honduras | 41 | 48 |
| Kenya | 61 | 54 |
| Pakistan | 26 | 27 |
| Uganda | 27 | 36 |
| Guatemala | 44 | 43 |
| Sri Lanka | 38 | 28 |
| Venezuela | 34 | 33 |
| Syria | 45 | 50 |
| Colombia | 37 | 33 |
| Jamaica | 21 | 32 |
| Chile | 36 | 39 |
| Tunisia | 43 | 35 |
| Argentina | 32 | 20 |
| Philippines | 42 | 43 |
| India | 22 | 7 |
| Iraq | — | — |
| Morocco | 58 | 65 |
| Indonesia | 51 | 58 |
| DEVELOPED COUNTRIES | | |
| U.S.A., (1972) | 8 | |
| U.K. (1970, 1972) (average) | 25 | |
| Germany (1972) | 16 | |

* Capital goods defined as S.I.T.C. 7 except 732·1, 661, 67, 68, 69.

SOURCE

UNIDO *Industrial Development Survey*, 1974.
U.N., *Commodity Trade Statistics*, 1970, 1972.
U.N., *Yearbook of National Accounts Statistics*, 1974.
U.N., *Statistical Yearbook*, 1973.

of technological dependence. On the whole developing countries import a much higher proportion than developed countries, as Table 5.2 indicates, though experience varies between countries. India and Brazil, both of which have built up capital goods capacity, have low ratios, while within the industrialised countries many of the small countries have relatively high ratios.[15] These import ratios are not a reliable indicator of the degree of technological dependence. For one thing, as argued in the next chapter, capital goods capacity is a necessary but not a sufficient condition of technological independence. Many countries produce capital goods locally on the basis of imported technology. Moreover, countries with substantial local technological capacity may none the less have a high import ratio exporting some capital goods and importing others. This probably accounts for the high import ratio among small industrial countries.

The pattern of trade in capital goods indicates the heavy dependence of poor countries on imports from the advanced countries. 89 per cent of South imports of machinery and transport equipment come from North countries, 6 per cent from centrally planned economies and only 5 per cent from other South countries.

### THE CONSEQUENCES OF TECHNOLOGICAL DEPENDENCE

The transfer of technology from advanced countries has enabled countries of the third world to benefit from the manifold developments of science and technology in the industrialised countries, during the past two hundred years. The transfer has permitted countries to use this technology without themselves going through the difficult and costly process of developing it. This is one of the main advantages of being a 'late-comer' in terms of development. Much has been written about the advantages thus conferred on third world countries, in discussion of foreign investment and of technology transfer.[16] Not only has technology transfer permitted the use of high-productivity techniques: it has also, in many cases, inspired the desire for technical change, which forms an essential basis of industrial development, whether based on imported or locally developed technology.[17]

In most of this chapter we shall be concentrating on the undesirable consequences of technological dependence. However, the 'late-comer' advantages of technology transfer form a background to the discussion. While the advantages are often not overriding, they may explain why a country may be justified in pursuing a policy of technology transfer, and hence permitting the associated technological dependence, despite the considerable costs so incurred.

The two advantages of technology transfer mentioned arise from *transfer* of technology, not from *dependence*, whilst the undesirable consequences arise from technological dependence, rather than transfer as such. That is to say the main disadvantages of technology to the third world arise from the fact that they are technologically dependent on the advanced countries, and receive their technology in a more or less one-way flow from them, not from exchange in technology as such, which may confer considerable advantages.

The undesirable consequences of technological dependence may usefully be classified into four categories:

(i) cost;
(ii) loss of control over decisions;
(iii) unsuitable characteristics of the technology received;
(iv) lack of effective indigenous, scientific and innovative capacity, which is itself a symptom of underdevelopment.

The four categories are interrelated, affecting and reinforcing each other. Each has consequences for the extent and pattern of development.

(i) *Cost*

It is extremely difficult to arrive at comprehensive figures for the cost of importing technology among developing countries. This is because the technology transfer takes so many divergent forms, paid for in divergent ways. Direct payments for technology, in the form of royalties, licence fees, etc., only cover a small proportion of the total payment, which should also include the element of technology payment included in imports of plant and machinery, and of goods, in payment of fees and salaries to qualified foreign personnel, and in the profits remitted on foreign investment. For tax and other reasons, payments for technology and remittances of profits may take the form of overinvoicing of imports and of underinvoicing of exports. The evidence suggests that the extent of this type of over and underinvoicing may be considerable.[18] The evidence collected has been concerned with payments involving overt technology imports. But no attempts have been made to quantify the technology element involved in payment for imports of plant and machinery where there is no overt technology import, nor with payments for foreign personnel.

There is a conceptual problem in identifying the total cost of technology payments. While we know that an element in many payments is a return for technology, it is difficult to establish the size of this element. A major problem is the large monopolistic or near-monopolistic factor in the market for technology, and associated flows of investment, goods and manpower. For all these flows the payments exacted may be more than the competitive norm, as established with reference to world prices in a more nearly competitive situation. And for many of these flows the actual division into different types of payment (profits versus royalties for example) may be arbitrary, reflecting tax and other advantages of a particular set of cash flows, and not costs, nor the sources of monopoly power. The monopolistic element in the market permits a type of quasi-rent (i.e. returns in excess of cost plus normal profit) to enter into the returns. The correct division of these quasi-rents needed in order to establish the cost of technology, depends on how far it is technology and how far other factors that are responsible for the monopoly element.

UNCTAD has made estimates of the *direct* costs involved in *overt* technology transfer.[19] The estimated costs of payments for patents, licences, know-how, trademarks, management and technical fees were \$1500 million in 1968, or nearly 0·5 of Gross Domestic Product, and 5 per cent of exports. These figures need revising upward to include underpayments through transfer pricing, and to include the cost for technology transferred implicitly via sales of product and the payment for foreign personnel, not involving overt technology contracts. Imports of machinery, equipment and chemicals cost third world countries \$18,420 million in 1968. Assuming one-tenth of this cost was a return on the technology involved in their production, this would more than double the cost of the technology transfer, bringing it up to 10 per cent of exports. Similarly, some proportion of the cost of imports of other manufactures may also be included as payment for technology, further raising the total cost.

UNCTAD has also estimated the likely rate of increase of overt technology payments: it is estimated that these payments will rise by around 20 per cent per annum to 1980, raising the total cost to \$9000 at that date, or 15 per cent of total exports. Assuming that the non-overt costs rise in line, then by 1980 as much as one-third of third world export receipts may be required as payments for technology.

These are the direct foreign exchange costs: there are also indirect costs, which take the form of restrictions on sources of input and access to market outlets.[20] These restrictions may be part of the formal agreement for the transfer of technology, or may, as with much direct foreign investment, be an informal and automatic result of operations being conducted as a subsidiary of a multinational company. Among the most common restriction is that of tying the purchase of imported inputs, equipment and spare parts to a particular source. Like tied aid this often has the effect of tying the purchase of imports to a more expensive source than would be used in the absence of the agreement, and thus of increasing the cost of the transfer. The other aspects of indirect costs arising from restrictions – e.g. restriction of exports, limitation of competing supplies, discouragement of use of local personnel – also increase the implicit cost of the transfer, but perhaps their most significant effect is in reducing local control. So far, though it has been established that restrictions of this type are widespread in technology contracts, the cost has not been established.

The heavy costs of technology transfer for third world countries are in large part due to their situation of technological dependence. In the first place, it is technological dependence that leads to the necessity for the net import of technology and the consequent net payments. Secondly, the situation of technological dependence is responsible for the very weak bargaining position of many developing countries *vis-à-vis* technology suppliers, and consequently for the poor terms exacted.

In the market for technology, bargaining power is of key importance in determining the terms of transfer. This is because the market for technology is, of its nature, imperfect. In a perfect market, competition would reduce the cost of acquiring technology to its marginal cost. But once the technology has been developed, the marginal cost is very low, sometimes approaching zero. The system of commercialisation of technology, consisting of legalised monopolistic practices, like the patent system, product differentiation and trademarks, has permitted the monopolisation or oligopolisation of the market for technology and hence the sale of technology at a price far in excess of its marginal cost.

Commercialisation has thus established a system whereby sellers of technology are able to exact a price in excess of the cost of its communication to buyers. The exact price depends on the maximum the buyers are prepared to pay for the technology, the minimum at which the sellers are prepared to sell it, and the bargaining strength of buyers and sellers.

The maximum price the buyers are prepared to pay depends on their estimation of the value of the technology, and the cost of alternative ways of acquiring the technology, including buying it from other sources or developing it themselves. Their estimates of these are often uncertain – particularly since it is the essence of buying technology that one does not know exactly what one is buying. Estimates are likely to be weaker, and more susceptible to sellers' influence, the weaker the country is technologically. This is one reason why technological dependence tends to lead to a bad bargain. The minimum price the seller is prepared to accept depends on the actual costs of imparting the information and the monetary loss he would incur by imparting it. The latter, which is normally of much greater importance than the costs of communication, consists in the loss in revenue consequent on the dilution of his monopoly power as a result of parting with the technology, and also the potential loss of revenue on sales of the same technology elsewhere in the world, which may result from reducing his price in any particular case. There is a sort of spiral effect which the seller takes into account, as the terms of a technology bargain in one part of the world influence the terms the seller receives elsewhere, and this spiral effect helps determine the minimum price he is prepared to accept in a particular case.

Thus, although the actual cost to the seller of imparting his information may be very low, this may bear little relation to the minimum price he is prepared to accept, because he also takes into account his potential monetary loss from striking a poor bargain. Similarly, although the cost of developing the technology again may be very high, the buyer of the technology is rarely concerned with this price, but more often with whether it is worth acquiring the technology at all, and with alternative sources. None the less, there is often a wide gap between sellers' minimum and buyers' maximum prices, and this is where bargaining becomes important.

Technological dependence makes a high price likely. First, it may rule out the possibility of reproducing the technology oneself. Even where this is a possibility, and indeed even where local technology has already been developed – see section (iv), below – the market structure and the prejudice that favour the use of foreign technology, which stem from technological dependence, tend to lead to an exaggeration of the benefits of foreign technology, and an underestimation of local technical capacity. This raises the price the buyer is prepared to pay for the foreign technology. Secondly, technological dependence reduces technical knowledge on the buyer's side, severely limiting the ability to search for alternative sources, and the ability to estimate realistically the gains from acquiring the technology. Thirdly, it reduces bargaining power in the sense that the buyer has little or nothing to withhold which the seller wants. In trade in technology between advanced countries cross-licence agreements are widespread; such agreements temper the exploitation of monopoly power on the part of the sellers.

Countries differ in their local technical capacity and in their bargaining power *vis-à-vis* technology sellers. They may also differ as between different industries. But in general it remains likely that those countries which are weakest in their own technical capacity are also most technologically dependent, and in the worst position to strike a satisfactory bargain with technology sellers. Comparisons of the terms reached with different countries, at different stages of development, would be instructive in this respect.

The final price agreed on, for any individual technology contract, is likely to be the more unfavourable to the buyer the more widespread technological dependence is. This is because, on the demand side, the more widespread this technological dependence the fewer are likely to be the known alternative technologies. And also, because from the point of view of the seller of technology, the greater the technological dependence among developing countries, the greater the quasi-rents he receives from his technology, and therefore the greater his potential loss should he strike a 'poor' bargain with any individual country, since this might lead to poor bargains elsewhere. Hence, given any degree of technological dependence in a particular country, the price agreed on for technology contracts is likely to be the higher (with both the minimum supply price and the maximum demand price higher) the more prevalent technological dependence is among developing countries generally. Similarly, the cost of acquiring technology in any country is likely to be reduced by a reduction of technological dependence in the world as a whole. Thus, developing countries stand to gain from reduced dependence elsewhere.

The process of arriving at a price for technology may be illustrated diagrammatically. Figure 5.1 below shows how the buyer's maximum price is arrived at. The triangular shaded area, $b$, $b^{i}$ and $b^{ii}$, shows the buyer's estimates of the value of technology to him. It is a triangular area, rather than a line to indicate the uncertainty with which the buyer regards the value of the technology. As will be seen, the area of uncertainty increases with the degree of technological dependence of the buyer, and the minimum value also increases, indicating the exaggerated value placed on foreign technology in accordance with the degree of technological dependence. The band $AA^*$ (also an area, not a line, because of uncertainty) indicates the estimated cost of reproducing the technology oneself. This rises sharply with the degree of technological dependence until it becomes vertical, indicating the impossibility of reproducing the technology. The third line on the diagram, $CC^{i}$, is the estimated cost of acquiring the technology elsewhere. This too rises with technological dependence since a country's ability to look around intelligently rises with its own scientific capacity. Its desire to shop around also tends to be a function of technological independence. The operative line, in technology bargaining, is the lowest point on the three curves, i.e. the buyer's maximum price, in Figure 5.1, is $ABC^{i}$, $AB$ representing the cost of reproducing the technology oneself, for the

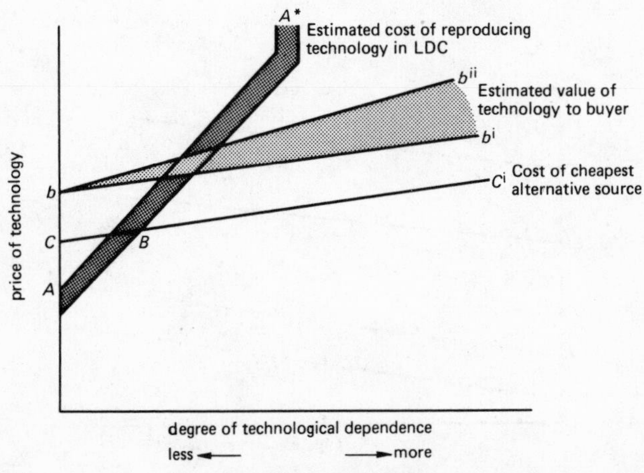


Fig. 5.1 Buyer's maximum price

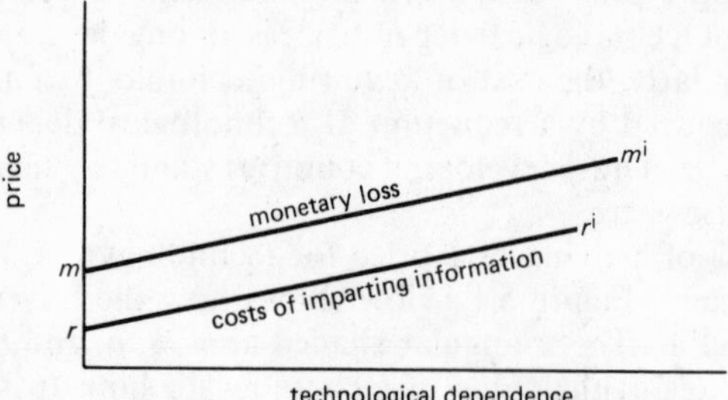


FIG. 5.2 Seller's minimum price

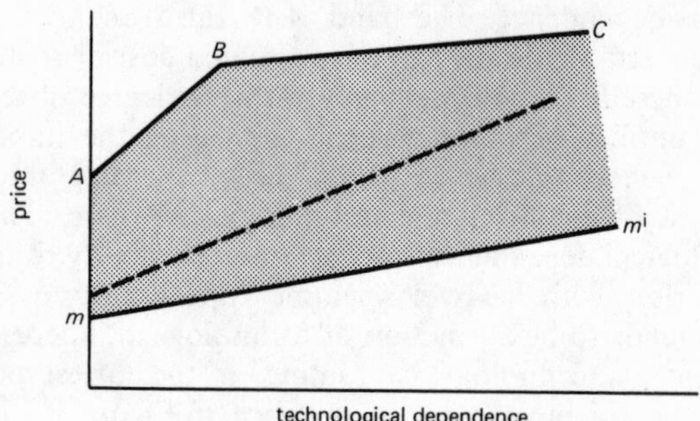


FIG. 5.3 Buyer's maximum v. seller's minimum

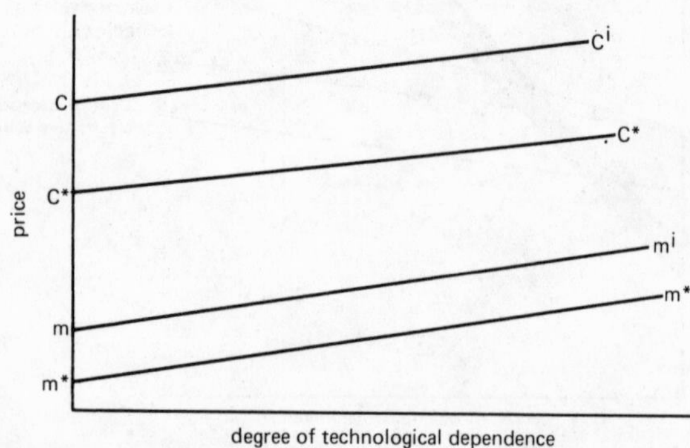


FIG. 5.4 Effects of overall technological dependence

least technologically dependent countries, and $BC^{i}$ the cost of alternative sources.

Figure 5.2 shows the determination of the sellers' minimum price. $rr^{i}$ is the cost of imparting the information. It rises with technological dependence because the more technologically dependent the country is the more general information it needs to use any particular technology, and the more skilled assistance it needs. The line $mm^{i}$ is the estimated monetary loss from imparting the technology due to (*a*) loss of sales using seller's technology elsewhere; and (*b*) the spiral effect on other technology bargains. It is assumed that this too rises with technological dependence because the seller recognises that the less technologically dependent the country is the more likely that it will go ahead with production, whether it acquires the seller's technology or not. Hence the seller is likely to face losses in sales in any case. The seller's actual minimum price is determined by the highest of the curves: in the diagram this is $mm^{i}$.

Figure 5.3 puts the two diagrams together. The shaded area represents the bargaining area. Within that area determination and skill determines price. Not only does the shaded area as a whole rise with technological dependence, but the likely price within that area is the more likely to be favourable to the seller, and unfavourable to the buyer, the greater the technological dependence, as shown by the broken line. This is because bargaining skill, *and* determination, are likely to be least for the technologically dependent, most for the independent. Countries which are highly dependent often lack independent negotiators, and rely on foreign personnel to bargain for them.

The three diagrams ignore the degree of dependence elsewhere. As technological dependence elsewhere decreases, so the costs of alternative technologies decrease, so line $CC^{i}$ in Figure 5.1 shifts downwards to $C^*C^*$ as shown in Figure 5.4 while the seller's monetary loss also decreases, i.e. line $mm^{i}$ shifts down to $m^*m^*$. Hence the whole bargaining area shifts downward for each country, if overall technological dependence is reduced.

(ii) *Lack of control*

Economic independence has been defined as a 'situation in which national institutions (including private business and interest groups) have the right, capacity and power to take and implement decisions affecting the national economy and its component units without a *de jure* or *de facto* veto power being held by foreign individuals, enterprises, interest groups or governments'.[21] Economic interdependence between individuals, countries and enterprises is such that complete independence, in this sense, is not a possibility. But there are big differences in the extent of national control over local decision making. At one extreme, with most assets in foreign ownership most of the economic decisions may be made outside the country. This was the situation which many countries faced in the early 1950s. A powerful motive for nationalisation was the desire to secure local economic control. Increasingly, new foreign investment was only permitted a minority shareholding in the ventures to which it contributed, another aspect of the attempt to secure economic control. However, it is becoming increasingly evident that ownership as such means very little so long as countries remain technologically dependent. Technological dependence has as an important consequence a severe dilution of economic control, *irrespective* of the nationality of the ownership of the assets.

Some aspects of an enterprise's operations are outside its control because they are the products of the decisions of other economic units: for example, the price and availability of raw materials may be of this nature. Decisions which normally are within the control of the individual enterprise are those on the quantity and nature of investment, price levels, quantities produced, suppliers and purchasers, the allocation of profits, etc. Technological dependence tends to remove these latter decisions from local control. The extent to which it does so depends on the extent, nature and form of technological dependence. Technological dependence may take the form of foreign-financed investment, using foreign managers; of joint ventures with some foreign managers and finance; and of local ownership with contracts to secure foreign technology.

Clearly foreign investment involves the most complete loss of control. Although countries can and do impose some restrictions on the activities of foreign investors, by its nature foreign investment removes control of all the main economic variables into foreign hands. Localisation of manpower requirements[22] may retain managerial office formally in local hands, but the main decisions are often taken in the head office of the company. Subsidiaries of multinational companies naturally have many of their decisions taken as part of the overall plan for the company as a whole. These include decisions on investment, pricing, profit remittances, sources of inputs and outlets for output. Joint ventures have been advocated as a means of avoiding some of this loss of control. Much depends on how large the foreign minority shareholding is, who the local partner is, and how determined the local partner is to take control. Subrahmanian concludes, after studying cases of joint venture in India: 'the general conclusion emerging from the analysis . . . is that foreign firms have managed to dilute ownership-mix in such a way as not to cause any significant diminution in foreign control in the individual cases.'[23] Joint ventures share some of the characteristics of foreign investment and some of those of local investment using foreign technology; the balance of the shared characteristics depends on the circumstances in the particular case.

While the consequences of both foreign investment and joint ventures can be attributed to technological dependence, since it is the need for foreign technology and for foreign management skills that provides their justification, the implications of technological dependence as such can, perhaps, be seen in their purest form in the case of totally locally-owned firms which buy their technology from abroad. Sercovich has concluded a study of such firms in the Argentine. He finds that the content of the technology agreements are such that most of the power of independent decision making is taken out of the hands of the local owners and managers. Licensing agreements for the import of foreign technology include clauses determining variables such as: source of inputs of machinery, materials, spare parts; prices and quantity of output; export outlets permitted; restraints on the dynamic effect of the transfer – e.g. non-transference of patents, restrictions on local R and D. Similar restrictions were found by Subrahmanian[24] in his study of Indian firms. For example, 23 per cent of the agreements he examined involved some restrictions on local sales (type, quantity, price) while over half involved restrictions on exports. Replies to an UNCTAD questionnaire[25] revealed how widespread such restrictive clauses are.

These clauses mean that many of the decisions normally thought to be part of managerial responsibility are taken outside the control of the local firm and outside the country. Technology contracts provide explicitly for the same sort of dilution of local control that happens automatically and without explicit provision in the case of foreign investment. Although Sercovich has argued that the basic decisions – over investment and allocation of profits – remain with the local firm, these decisions too are very much influenced, if not totally determined, by the decisions taken which are beyond local control. Since ownership alone does not appear to secure much control, resources used to secure ownership might be better devoted to technological development, if the ultimate aim is control over a nation's own assets.

The loss of control resulting from technological dependence weakens the bargaining power of the local firms and hence increases the cost of imported technology. It also makes it more likely that the imported technology will be associated with unsuitable characteristics. By inhibiting local R and D and tying the local firm to further technological developments overseas, the technology contracts often prevent or at least inhibit the local adaptation of the imported technology.

(iii) *Unsuitable characteristics*

Perhaps the most obvious feature of imported technology is that its design is normally originally intended for the *producing* country, and hence its characteristics may be unsuitable for the *importing* country. In trade in technology between rich countries and poor countries unsuitable characteristics are particularly likely to be the consequence. The inappropriate nature of much technology designed in advanced countries formed the main subject of Chapter 3. The discussion will not be repeated here. But it must be emphasised that this is an important (perhaps the most important) consequence of technological dependence, where countries rely on imported technology and therefore do not have technology designed for their own conditions and needs. The manifold consequences – on employment, income distribution, etc. – following from the use of inappropriate technology, discussed in Chapter 3, are also aspects of technological dependence of the poor countries on the advanced countries.

(iv) *Lack of effective indigenous scientific and innovative capacity*
Technological dependence, of course, is largely the result of the lack of local technology; but it also contributes to the lack of an effective local scientific and technical capacity. It does so in two ways: by inhibiting the process of learning-by-doing in technical development, which is essential for the development of scientific capacity; and by leading to a structure of productive activity, which tends to make the activities of the local scientific and technical institutions either totally irrelevant, or poor images of advanced-country institutions.

Many of the skills that are needed for innovation, for installing new technologies and in the operation of industrial plant, can only be acquired through learning-by-doing. The development of skills in plant construction is a good example. The first time that an engineering firm carries out a plant construction contract, it quite often runs into unforeseen problems, which increase capital costs by lengthening the period of construction, or which result in a waste of materials, or cause inefficiencies in the operation of the plant itself. The contractor is much less likely to run into similar problems the next time he builds a similar plant; with each successive plant, costs of construction fall and construction time gets shorter. In technologically complex industries, like chemicals, the maximum efficiency of construction is only reached after the contractor has done five or six plants.

There are similar learning effects in other scientific and technological activities related to production. The skills that are needed to build minor modifications into installed plant also can only be acquired by doing.

So, if developing countries are to build up the human skills and the institutional systems that are needed to reduce the extent of technological dependence, firms, laboratories and engineering organisations must have opportunities for learning-by-doing. There has to be some way of coping with the short-run costs of inefficiency and inexperience in their activities in order to benefit from the long-term gains of cutting down dependence on foreign skills and technology.

The trouble is that it is in the nature of technological dependence, as we have described it, that it seems to prevent this process of learning by-doing from happening – or at least drastically cuts down the scope for it. Foreign licensors or direct investors very often insist on using their usual machine suppliers in the advanced countries; they contract advanced-country engineering designers; they may have exclusive agreements with advanced-country plant contractors, who are then brought in to put up the plant in the developing country. Even if the foreign supplier of process technology is working with a locally-owned firm (through a licence or a joint venture for example) there will not necessarily be any pressure for it to use local skills rather than foreign. Local businessmen also often have strong preferences for foreign engineering firms and contractors – precisely because they also are suspicious of the inexperience of local engineers. In a similar way, local businessmen are often unwilling to use local *technologies* even where they are available, because foreign technologies which are commercially proven, and which are supplied along with experienced technicians and engineers from the foreign supplying firm, are a much less risky proposition. So there are repeatedly cases where local scientific and engineering laboratories in developing countries have been able to develop a technology to the point of commercial production – only to find that local firms prefer to license a precisely similar technique from abroad, in spite of higher financial costs.

## POLICIES TOWARDS TECHNOLOGICAL DEPENDENCE

Two strategies may be distinguished: first, one of *controlling* technology imports, so that the undesirable consequences of technological dependence are modified or offset; secondly, that of making a direct attempt to *reduce* the extent of technology imports. Both policies are made difficult, if not impossible, by the general relationship of dependence of the third world on the advanced countries. Discussion of policies thus reveals a close connection between the somewhat pragmatic view of technological dependence, discussed above, and dependency theories.

### *Controlling foreign technology*

During the past half-century two countries succeeded in basing their industrial development on imported technology, and used this technology, successfully, as a basis for their own technological development: these two countries are Japan and Russia. Both show that it is possible to import technology and yet not be swamped by it; lessons from their experience are therefore of particular relevance.[26] These two countries have had three things in common in their policy towards technology imports. First, they maintained a strict control over technology imports and deliberately restricted them to particular areas where they considered the need greatest, and where they would not inhibit local development efforts. Secondly, they allowed technology imports via licensing agreements, but they did not allow (except for special cases in Japan) the imports to be accompanied by foreign investment, with majority foreign shareholding. Thirdly, they adapted the technology they received, and rarely introduced it unmodified. In Japan one-third of total R and D expenditure has been devoted to adapting foreign technology, and the average expenditure devoted to adapting a unit of technology has been greater than the average expenditure devoted to creating a unit of local technology.

The evidence, such as it is, suggests that both Japan and Russia were successful in offsetting or modifying much of the deleterious effects of the import of foreign technology. In a way both these countries *appear* to illustrate the truth of Gerschenkron's thesis on the advantage of being a late-comer, having apparently used developments abroad to enable them to skip aspects of the development process of the first-comers. Yet, the examples of both could be used to question the thesis, because they suggest that *uncontrolled* technology transfer from more advanced countries may be disadvantageous, rather than advantageous. The technology policy in both countries modified the consequences in each of the categories considered here. In Japan, Ozawa shows that through a tough policy the terms of technology imports were reduced from a typical 8 per cent royalty, with a contract duration of fifteen years, to 3–5 per cent lasting five years: 'with the growth of the Japanese economy and the increasing availability of technology, both in Japan and abroad, the bargaining balance shifted from western licensors to Japanese licensees.'[27] Multinational companies, having previously been keen to supply technology under licence, pressed to be allowed to invest directly to secure better returns. The foreign licensors do not appear to have secured successful control, certainly not over exports, nor over the use of the technology as the basis for local R and D. Ozawa has shown that the most successful export industries were those which were heavy technology importers. Granick shows how imported technology was much more successful when adapted by local scientists and technologists to suit local conditions.[28] He believes that the different conditions necessitated technological adaptation: 'even in those factories which were most directly based on Western technology, and in which aid from foreign companies was greatest, the need for small changes and innovations culminated in the necessity for a massive technical effort.'[29] The heavy R and D expenditure on imported technology in Japan shows that the technology did not inhibit but stimulated local science and technology. In 1962 Japan spent nearly 1 per cent of GNP on R and D, while industry financed almost the total involved, indicating that the expenditure fulfilled local economic needs. In most third world countries, total expenditure as a proportion of GNP is far lower than this and industry only finances a miniscule proportion.

The experience of Japan and Russia then suggests that a controlled policy towards technology imports can succeed in securing many of the advantages of technology imports while avoiding the worst consequences. However, other countries have set up a similar structure, without securing the same results. India is the obvious example. On paper her policy towards science and technology imports allows only selective imports; she avoids foreign investment wherever possible and negotiates toughly on terms. Yet practice has not mirrored policy. Foreign technology imports have inhibited local science and technology; control has been lost to the foreign licensors; local R and D does not make use of and adapt foreign technology but, on the whole, imitates it. Closer study of all these countries is needed to provide full backing for these claims, and a soundly based explanation. But, tentatively, the paradox may be explained in terms of stage of development and structure of production. Both Japan and Russia were advanced but backward countries, which were in a much better position than India to develop rather than be swamped by foreign technology; and both had rigid controls against the outside world. In contrast, despite the apparatus of import controls, India has close contacts with the advanced countries and until recently has encouraged the import of technology. Vested interests have developed in the production of Western-style products using Western-style technology, and the technology policy is not tough enough to stand up to the interests it would need to counteract to be successful. To be successful, a technology policy must succeed in breaking the ties with advanced-country companies, tastes and products, which technological dependence has built up. This sort of break is the more difficult to achieve the more integrated the ties and the interests that have developed in the system; on the other hand, while in one way it is easier for less developed and less integrated countries to break the ties at an earlier stage of development, it is more difficult in another because they lack the technological capacity to make the break.

*Reducing dependence*

The dichotomy posed between controlling foreign technology and reducing dependence may be a false one. Successful control of technology will have, as an important long-term consequence, reduced dependence, as local scientific capacity is developed on the basis of the imports.

There are some differences in policies between the two strategies. Policies of modifying technology imports are concerned to use foreign technology as the lever or take-off point for local technology, whereas policies aimed at reducing technological independence would be more concerned to develop local technology as an *alternative* to foreign technology, i.e. to replace foreign technology entirely. This could mean a quite different distribution of imported technology. While modifying foreign technology would tend to distribute local scientific resources more or less *pari passu* with imported technology, so as to adapt and modify and learn from it, a policy of creating a local alternative technology would allow foreign technology only in those areas where a local alternative was out of the question. Modifying foreign technology would make local and foreign R and D complementary in the same industries; a policy of reducing dependence would make them complementary but in different industries. The modifying policy would tend to be less effective than a direct attack on dependence, in terms of reducing overall dependence, because the complementarity of local and foreign R and D would normally require the continued import of foreign technology. Much depends on whether foreign technology really can be controlled in an ideal way, such that local adaptation and learning occurs, eventually to be transformed into local technological independence. This ideal policy sounds very good on paper, making the best of foreign technology without its ill-effects. In practice it may well be an impossible policy for most countries. This is because technology imports are not policy-neutral, so that they can be encouraged/discouraged, selected/rejected in a pre-planned way; once allowed, they establish a hold over policies by changing market and economic conditions, and creating vested interests, which makes them extremely difficult to manage, and indeed may largely reduce the desire to manage them. Perhaps they might be likened to a drug: with addictive drugs, a plan to use them to stimulate the nervous system and incite a feeling of general well-being, and then gradually stop their use as the system moves to a higher level of well-being without them, sounds excellent – but is extremely difficult to put into practice. The question then is whether technological imports may be managed and are non-addictive, as perhaps the experience of Japan might suggest,[30] or are addictive and unmanageable, as the experience of India might argue. In the former case, a policy of modifying the consequences by a controlled strategy followed by a gradual move towards independence is a possibility. In the latter it is not; and if reduced dependence is an aim, then it must be pursued as such, and not indirectly via controlling the consequences of importing technology. From this point of view, India's policy of developing substitute technology is rational: the only drawback is that it has not been *used* as a substitute in practice; rather the foreign technology has been the substitute. But what is required then is far more extensive prohibition of the use of foreign technology, rather than redistributing scientific and technical resources to make them complementary with foreign imports. Again, however, the possibility of this as a realistic policy depends on how far the foreign technology and the interests to which it gives rise have penetrated in the power structure.

In terms of actual policies, a direct attack on technological dependence would require the minimum import of foreign technology, and the maximum encouragement and use of local technology. How far this 'minimum' or 'maximum' actually took one would depend in part on local technological capacity, and in part on what cost the country was prepared to bear in the short run to achieve independence. There are bound to be heavy short-run costs, in terms of loss of output and incomes, as a country struggles to manage on its own.

It is clear that such a policy is more nearly a possibility for some countries than others. Small countries with neither technicians, experience, nor capital goods capacity are in no position to aim for technological independence, whereas some larger countries have considerable local capacity. This brings us to the question of why *national* independence should, in any case, be an aim. National economic or technological independence has little rationale in terms of the independence of homogeneous interests; nor is it justified from the point of view of economic efficiency. Many national units are far too small to provide a sensible basis for economic or technical autarky. For each nation, big or small, to duplicate the economic and technological efforts of every other is as irrational and inefficient as the economic arrangements of the Zollverein states.

Many of the disadvantages of dependence *vis-à-vis* advanced countries could be eliminated, without incurring the heavy costs of autarky, by greater specialisation and exchange *between third world countries.* Exchange between third world countries would avoid the unequal exchange[31] associated with exchange in technology between advanced and third world countries; it would allow for the development of a more appropriate technology for the third world, and for specialisation in its production; and it would remove the inhibiting effects of advanced technology on learning. Where the import of advanced technology is required, by negotiating jointly third world countries could much increase their bargaining power, reduce the extent and hence cost of multiple collaboration through the third world as a whole, and hence reduce the costs of importing the technology.

If a third world technology policy is to get anywhere a prime need is for institutional machinery to establish technical links between third world countries, exchange information on technical know-how and capacity and provide for bargaining assistance and eventually for joint negotiation. At the moment, as with most other areas of policy, all links are North/South, not South/South, so that innovations are rarely transferred between third world countries, and nearly always from advanced countries, even where third world countries have suitable technology available. This is partly because of historical ties between advanced countries and ex-colonies; partly because financial ties are almost all in that direction; and partly because the same sort of prejudice that thwarts the use of local technology where foreign is available also acts in favour of developed country technology and against technology from other underdeveloped countries.[32]

Technological dependence can be seen as cause and effect of the general dependency relationship briefly described in the first part of this chapter. It is cause in so far as the need to import technology – lacking an indigenous technological base – leads to foreign investment, loss of control and the introduction of advanced-country patterns of consumption and production. An enclave economy, dependent on the advanced countries, and with its main links for inputs, markets, management, finance and technology with the advanced countries, then develops. The situation is self-reinforcing because once advanced-country technology has been introduced it creates a society in its own image, requiring further import of technology to feed the markets which have been created, and to enable the industries to survive and expand. Given a productive structure based on the production of advanced-country products, using advanced-country techniques, the natural consequence is that the local science and technology systems are small and irrelevant, adept at assimilating (unadapted) foreign technology, but lacking independent innovatory force. Yet the weakness of the local scientific/technological base is not only an outcome but also a prime cause of technological dependence, and indeed of dependence generally, because it means that there is no real alternative to the import of foreign technology. There is a vicious circle in which weak technology reinforces dependence, and dependence creates weakness.

Attempts to break out of the cycle tend to be thwarted by the attitudes and interests developed as a result of the dependent relationship. Action on science and technology alone, and on the *terms* of transfer of technology, are likely to be ineffective without more general action on economic dependence, because the eco-political structure resulting from this dependence requires further import of advanced-country technology. But attempts to reorient the whole economy away from the dependent relationship are prevented by the loss in efficiency that would result, in the absence of an effective alternative technology, *and* by local interests that have developed in the continuation of the system. These obstacles were apparent in the policy discussion above. The dependent relationship means that the advanced countries' interests are internalised, inhibiting independent action to counter technological dependence. One possibility is that third world countries combined might counter these interests, and create new ones in an alternative direction. The Andean pact countries have taken the first steps in this direction.

NOTES

1. See Furtado (1964), Frank (1964), Sunkel (1972), Dos Santos (1973), Szentes (1971), Amin (1973), Griffin (1969), who may all *loosely* be described as part of the dependency school. Indeed the term is somewhat ill-defined and might be used to include many writers outside the neo-classical tradition.
2. The distinction between the Marxist school (e.g. Baran (1957), Szentes) and the 'structuralist' school (e.g. Frank, Furtado, Griffin) is brought out most clearly in an (unpublished) paper by Sanjaya Lall. I am indebted to him for this and for

other ideas. See S. Lall, 'Is "Dependence" a Useful Concept', *World Development*, Vol. 3 (November/December 1975).

3. See particularly Furtado and Frank.
4. The term 'commercialisation of technology' was introduced by Vaitsos.
5. O. Sunkel (1973) has produced an illuminating diagrammatic presentation of this situation, showing the close relationship between advanced countries' industrial systems and the formal sector of third world countries, with consequent marginalisation of the informal sector.
6. The study for UNCTAD by C. Cooper and F. Sercovich provides a useful description and analysis of many of these methods (see especially Part Two). It should, however, be noted that their study is deliberately confined to a rather narrow definition of technology transfer. It is concerned only with the formal sector, and within that sector with transfers 'required to set up or operate *particular* new production facilities'. It thus tends to exclude infrastructural technology and informal technology transmission: i.e. it excludes 'the more or less informal and unspecific ways in which productive enterprises and individuals in developing countries may learn about new technologies' (para. 35).
7. Cooper and Sercovich, para. 35.
8. See T. Ozawa (1971).
9. An important question for further research would be a detailed comparison of the terms on which advanced and third world countries get similar technology.
10. See, e.g., A. Bhalla's discussion (1974) of the effects of the cut-off of Russian technical assistance on technical innovations in the construction industry in China.
11. Cooper and Sercovich, para. 6.
12. K. K. Subrahmanian (1972). An Indian Chemical Manufacturers' Association Survey, *Preliminary Survey on the States of Research and Development in Chemical Industry* (Calcutta, 1970) showed that foreign-controlled firms did substantially less R and D than Indian-controlled firms.
13. Ozawa (1971).
14. See *The Role of the Patent System in the Transfer of Technology to Developing Countries* (U.N. publication, sales no. E75 II D6), Tables 7 and 12.
15. Maizels (1963) pp. 266–7 shows that in 1957–9 the large industrial countries imported on average 7 per cent of their gross investment in machinery and transport equipment; the small industrialised countries imported 59 per cent; the ratio for semi-industrialised countries varied between 21 and 100 per cent, and that for non-industrial countries between 59 per cent and 100 per cent.
16. See, e.g., D. L. Spencer (1970).
17. J. S. Mill emphasised this as one of the major advantages emanating from contact between nations, though he was concerned mainly with trade rather than technology transfer. To regard this as an advantage, of course, assumes that industrial development is regarded as a desirable goal.
18. See, e.g., C. V. Vaitsos (1974).
19. Some UNCTAD estimates are in TD/106, *Transfer of Technology*, especially Chapter 2. Similar estimates, based on the results of the UNCTAD questionnaire, are in TD/B/AC.11/10, *Major Issues Arising from the Transfer of Technology to Developing Countries*, especially Chapter III and Add.1, containing the country answers to the questionnaires.
20. See *Major Issues . . .*, Chapter IV, and especially Table IV–I. For a description of restrictions imposed in India see Subrahmanian (1972) Chapter VI, and for Argentina, F. Sercovich (1974).
21. R. H. Green, in D. Ghai (ed.) (1973).
22. See, e.g., Kenyanisation laws.
23. Subrahmanian (1972), p. 87.
24. Subrahmanian, pp. 143–63.

25. Reported in *Major Issues*, Chapter IV.
26. The discussion on Russia relies heavily on Granick (1967), while that on Japan relies on Ozawa (1966, 1971).
27. Ozawa (1966), p. 100
28. See the comparison between two tractor factories at Kharkov and Stalingrad established in 1930 and 1931. The Stalingrad factory was based on more or less unadapted U.S. technology, and made heavy use of U.S. training and U.S. personnel. The Kharkov plant substituted Russian for U.S. plant and personnel and contained considerable adaptations to the technology. Time of construction was substantially less and output per worker and per unit of investment higher in the adapted factory (Granick p. 118–19).
29. Granick, p. 117.
30. Though indeed Japan continues to rely heavily on the import of technology, and her way of life has been transformed by it in an apparently uncontrollable and not altogether desirable way; so it does not clearly demonstrate the non-addictive quality of the drug.
31. The concept of *unequal exchange* was used by Emmanuel to describe trade in *goods*; it applies to a greater extent to trade in technology.
32. This discussion about trade in technology also applies to trade in goods. A parallel case for third world trade in goods is considered in Chapter 7.

# [illegible] ›ods in Developing [illegible]

[illegible] ›stry has been a subject of considerable controvers[illegible] e[1] – see the Indian and Soviet debates – and in the[illegible] :onomists distinguish between *projects* rather than *indus*[illegible] to say their methodology tends to assume that no particular [illegible] lemerit) may be attached to a project because it is in a particular [illegible] y – rather each project is to be evaluated for the effects to be attri[illegible] to it. In contrast, Marxist tradition has been to make a sharp disti[illegible] ion between capital goods industry (Department I)[2] and consumer goo[illegible]s (Department II), and much analysis hangs on this distinction. Developments of this distinction – in particular those of Feldman[3] and Mahalanobis respectively – provided the justification for the build-up of heavy industry in Russia in the 1920s and 1930s and in India in the 1950s and 1960s. This chapter is concerned to explore these differences in approach, and to suggest other considerations, particularly technological development, which may justify special treatment for capital goods industries in developing countries.

It is tempting to spend considerable time on definitions, and impossible to avoid the question altogether. After all, definition and measurement of the capital stock has raised the major stumbling block to the production function approach in macro-economic analysis. In that debate objections to measurement of the capital stock are chiefly concerned with the problems involved in aggregating a collection of items which are heterogeneous within and particularly *over* time. Questions of *aggregation* need not worry us here – we require rather the possibility of *classification*. Those involved in the debate for the most part accept the possibility of classifying goods into capital and consumer goods. But even here there are major problems. While Joan Robinson accepts that capital goods may be identified – a 'Who's Who of individual goods' – she suggests that *all* goods in existence should count as capital goods, whether owned, or intended for, industrial use or for private consumers, or conventionally classified as capital or consumer goods. Two possible criteria for distinguishing between capital and consumer goods both turn out to be rather weak on closer examination: first, such goods may be distinguished on the grounds that capital goods render further services, while consumer goods are consumed 'instantly' – but this makes the classification dependent on the period of time taken to *constitute* 'instantly'; if interpreted literally it makes all goods capital goods, à la Robinson, and only services, which by their nature are consumed instantly, consumer goods.[4] A more generous interpretation of 'instantly' still requires reclassification of some consumer durables as capital goods, and some short-lived capital goods as either intermediate goods or consumer goods. An alternative way of viewing the question is to define capital goods as goods which are not demanded for themselves but as inputs which, together with other inputs, render further production possible. The significance of this as a method of classification depends on exactly how 'further production' is defined. Mrs Smith's coat would be a capital good if the service of keeping Mrs Smith warm in winters to come were to count as 'further production'. Apart from that, it would be difficult to avoid classifying much of workers' consumption as capital goods since by keeping the workforce alive and healthy it renders further production possible.

In view of these and other difficulties it is tempting to conclude that the attempt to distinguish between investment goods and consumption goods is misplaced and hence, of course, the main question under discussion – whether special efforts to build up the capital goods industry are justified – is a nonsense question. The impossibility of arriving at a sensible method of classification itself provides a very good reason for not discriminating, since discrimination requires classification. However, to accept this view would involve avoiding a real and not a nonsense question. While it may be impossible to achieve an overall watertight definition, it is surely useful to ask whether one should devote current resources to expanding production of steel or food, to production of machines which make machines, or machines which make textiles, and this is what the whole debate is about. Thus it will be assumed in what follows, as it is by the authors under discussion, that the distinction between capital goods (I-goods, or Department I goods) and consumer goods (C-goods, or Department II goods) is unproblematic. Broadly, we shall follow the same kind of classification as adopted by national income statisticians, which is also much the same as that used by Marx:

> The total product and therefore the total production of society may be divided into two major departments:
>
> I. *Means of production*, commodities having a form in which they must, or at least may, pass into productive consumption.
>
> II. *Articles of consumption*, commodities having a form in which they pass into the individual consumption of the capitalist and the working class.[5]

The Harrod-Domar identity,

$$g = \frac{s}{v}$$

where $g$ is the growth rate,
$s$ the savings ratio,
and
$v$ the capital output ratio,

provides a good starting point in looking at the role of the capital goods industry.

Given a constant ICOR (incremental capital output ratio), the growth rate is proportionate to the savings ratio. One may look at the determinants of the savings ratio in this identity in four ways: from the point of view of *savings capacity* in the economy – or the extent to which current consumption may be reduced, releasing resources for investment. Secondly, from the point of view of *investment capacity* – or the availability of the investment goods necessary to enable investment to take place. Thirdly, from the point of view of *absorptive capacity* – or the availability of projects. Shortage of managerial and administrative capacity, lack of skills, absence of required infrastructure and the need to do much investment sequentially rather than simultaneously may limit the number of projects possible at any point of time, in many LDCs. This limit is described here as constituting an absorptive capacity limit on investment. In practice it is likely that, rather than a sudden and complete ceiling, absorptive capacity limitations sharply reduce the returns on investment projects (or raise $v$ for any extra $s$). We shall return to this point later. Finally, in a capitalist economy in which investment decisions are made privately, *willingness to invest* is another determinant of the investment rate. Willingness to invest is related to absorptive capacity, since lack of absorptive capacity has the effect of reducing the returns to investment and obviously, therefore, willingness to invest. But in addition to technical conditions presented by absorptive capacity, business psychology and expectations and government incentives also help to determine willingness to invest.

For *ex post* investment to take place all four types of capacity must be present – that is, sufficient resources must be released and investment goods available, there must be projects in which to invest and a willingness to invest in these projects on the part of decision makers. In a socialist economy one might wish to emphasise the first three types of capacity and possibly exclude the fourth category altogether as an independent factor. In a mixed or capitalist economy willingness to invest on the part of decision makers is an important determinant of the level of investment. While *ex post* all four (or three) types of capacity must be present (and equal to the actual level of *ex post* investment), *ex ante* the potential investment capacity, in terms of these four categories, may differ. The level of investment and the subsequent course of the economy depends on the relationship between the *ex ante* capacities and the *ex post* reality.[6]

The maximum possible level of investment is set by the lowest of the four capacities. Assuming a constant and independently determined capital output ratio, then the maximum growth rate is also determined by the lowest of the four capacities. Different approaches to problems of development may often be reduced to different assumptions about which type of capacity provides the constraint on the level of investment. The starting-point and distinguishing characteristic of the Feldman/Domar and Mahalanobis models (for brevity referred to as the Feldman model in what follows) is that the investment capacity, or availability of investment goods, determines the level of investment. The model is one of a closed economy, so that it is also assumed that the availability of investment goods is determined by the output capacity of the domestic capital goods industries.

Starting with a given capacity to produce I-goods, the initial savings rate is determined by that capacity. But subsequent capacity to produce I-goods will depend on how far the initial investment is devoted to expansion of the capacity to produce C-goods, and how far to expansions of I-good capacity. Let us assume that $\lambda$ represents the proportion of initial investment devoted to investment in the I-good sector, and $(1 - \lambda)$ the proportion devoted to investment in the C-sector. The higher $\lambda$ is, the faster the expansion of the I-good sector, and the greater the production of I-goods possible in later years. With a constant ICOR (the same in both sectors), the higher $\lambda$ is, the higher the ultimate growth of the economy because $s$ is higher. While growth in consumption will initially be lower, the higher $\lambda$ is, eventually growth in C-goods capacity will speed up, as the larger size of the I-sector compensates for the smaller proportion of its output going to investment in the C-sector.

Suppose $Mt$ represents the initial capacity output of the I-sector, and $Ct$ the initial capacity output of the C-sector, so that $Yt = Mt + Ct$. The initial maximum savings ratio possible is then determined as $Mt/Yt$. Subsequently, investment capacity depends on the proportion of investment going to the I-good sector, since

$$M_{t+1} = I_{t+1} = S_{t+1} = M_t + \frac{\lambda}{v} M_t$$

Growth of the investment sector, in the first period,

$$\frac{M_{t+1}}{M_t} = \left(1 + \frac{\lambda}{v}\right)$$

The level of investment possible in the $n$th period,

$$In = Sn = M_t e^{\frac{\lambda}{v}n}$$

The consumption level depends on the initial C-goods capacity, and subsequent additions to that capacity, or

$$C_{t+1} = C_t + \left(\frac{1-\lambda}{v}\right)M_t$$

in the $n$th period,[7]

$$C_n = C_t + M_t\left(\frac{1-\lambda}{\lambda}\right)\left(e^{\frac{\lambda n}{v}} - 1\right)$$

The rate of growth of investment is thus always determined by $\lambda$ assuming $v$ is given and invariable. In the long run, the exponential term dominates in determining consumption, so that the rate of growth of consumption and of income is also, in the long run, positively related to the proportion of investment going to the I-sector.

This, crudely summarised, is the case for building up the capital goods sector, as presented by Feldman and others, the justification for the concentration on expanding I-goods at the expense, in the short run, of light industry capacity in both Russia and India. In Russia the policy did appear to lead, in the end, to a higher overall rate of growth, including a higher rate of growth of consumption output. In India perhaps it is fair to conclude that it did not.[8] The applicability of some of the assumptions on which the model is based may in part explain this varying experience.

The Feldman model is subject to two types of criticism: first for assuming that the operative constraint, and consequently determinant of the investment rate, is domestic investment goods capacity and not one of the other types of capacity discussed earlier. In so far as domestic investment goods capacity does provide the operative constraint then there is a clear case for building up the local capital goods industry. But if one of the other types of capacity provides the operative constraint then building up the local capital goods industries will result in unused resources. It can be argued that this is what has happened in India. In what follows therefore we shall be concerned to discuss the kind of conditions in which it is reasonable to expect local I-goods capacity to provide the constraint. Secondly, the model, along with many others, might be described as a *bottleneck* model. As such it is subject to the criticisms which are generally levelled at *any* bottleneck model. Indeed many of the criticisms of the model are specific applications of these general criticisms.

Bottleneck models, as the name implies, assume that a particular resource is available abundantly at a constant cost up to a certain limit, and then suddenly the supply is completely exhausted and no more is available at any cost. In the Feldman model this is the assumption made about the output of investment goods. It is also assumed that other resources cannot substitute for the resource in question, so that once exhausted in quantity, a total limit is imposed on all activity. The assumptions behind such models, therefore, are in complete contradiction to the assumptions of substitutability, diminishing returns and continuity that are at the heart of neo-classical economics. If we make these neo-classical assumptions then no single resource, such as I-goods capacity, can limit the total level of output. Domestic resources in the C-goods industry may move into the I-goods sector, foreign resources may substitute for local resources. Similarly, with other postulated bottlenecks – e.g. absorptive capacity – resources from elsewhere may move in to release the bottleneck. In practice therefore there can be no bottleneck sectors; rather there are diminishing returns as output of each sector increases and more costly and less efficient resources have to be drawn in. The pure neo-classicist would reject any bottleneck model for these reasons. Any theory which picks out one sector for special treatment – be it the capital goods sector, or the agricultural sector, energy or industry – is thus denying (often implicitly) that the assumptions of substitutability and continuity are applicable. Given some degree of substitutability then a project approach – using prices that reflect the degree of substitutability – becomes appropriate rather than a sector approach. Applying this to the Feldman model it would be argued that it is wrong to assume that investment goods are abundantly available up to a point, and thereafter not available at all – even in a closed economy. Rather investment goods would become gradually more expensive (and/or less efficient depending on which way you look at it) as the quantity of investment increases. This sort of criticism does strike at the heart of the model – indeed it means that the concept of 'investment capacity' loses meaning. It also means that it is wrong to assume $v$ constant, and hence growth uniquely determined by the savings rate. As savings rise $v$ rises continuously and this must be taken into account in determining the optimal savings rate. In a neo-classical model all constraints operate continuously; as with investment goods, similarly with savings capacity. There is no point at which savings are abundantly available at constant cost, nor a point at which savings suddenly 'run out'; rather they are continuously expandable at increasing cost. The same goes for the other types of capacity discussed. Absorptive capacity does not present a potential bottleneck – there are always some possible projects, but prospective returns are continuously reduced as investment increases.

Reality is probably more complex than either the pure neo-classicist or the bottlenecker would allow. While there may be some substitutability and elasticity of supply, it is often reasonable to postulate steeply rising costs and falling elasticity, so that the bottleneck model may provide a good approximation to reality, without being totally accurate.[9] On the other hand, some types of capacity fit better into the neo-classical framework than the bottleneck framework. In particular it might be more accurate to view absorptive capacity as causing a (possibly sudden) fall in returns rather than a complete bottleneck.[10] This would mean that the constant $v$ assumption of the model had to be dropped; as the savings rate rose $v$ might rise so that the maximum $s$ would not necessarily lead to maximum growth.

The other type of criticism of the model accepts a bottleneck approach but argues that, for most LDCs, the wrong bottleneck has been selected. The key assumption in the Feldman model is the assumed identity of investment with the capacity output of the I-goods sector, or $It = Mt$. Even assuming that it is the availability of investment goods that provides the operative constraint, only in a closed economy can this be identified with the output of the domestic I-goods industry. In an open economy I-goods may also be imported. Thus the *ex post* identity, $Id \equiv Sd$, for a closed economy, becomes $Id + If \equiv Sd + Sf$. While in a closed economy total investment cannot exceed – though it may fall below – total investment-goods capacity, in an open economy I-goods may be imported, and consequently even with zero I-goods capacity all the savings capacity potential of the economy may be realised by importing I-goods. In terms of the identity above, total savings and investment ($Sd + Sf$) can and is likely to exceed domestic I-goods capacity, the difference being made up by imported I-goods. In fact most LDCs do import a large proportion of their I-goods, so their domestic I-goods capacity is hardly relevant to the amount of investment they may do. Some part of investment consists of non-importable goods such as energy and construction: the model might therefore be applicable to these industries rather than heavy industry as a whole. But since I-goods to build up these industries may be imported, the model would need substantial modification. In an open economy, an upper limit to possible investment is imposed not by domestic I-capacity, but by that capacity *plus* foreign exchange available to buy I-goods from abroad, i.e. $Mt + F$, which sets the upper limit to investment. Assuming zero local I-goods capacity, then foreign exchange availability provides the upper constraint on possible investment. If this constraint is reached before the other savings constraints, the savings and investment ratio, and the growth rate, are determined by foreign exchange availability. This, of course, is what occurs in the well-known two-gap model of Chenery and others.

The Raj/Sen model shows how a rigid foreign exchange restraint of this kind may justify the build-up of heavy industry. If the limited foreign exchange is used to import consumer goods, the economy will not grow at all. If it is used to import investment goods which produce consumer goods, there will be a steady rate (or level, depending on the nature of the foreign exchange restriction) of investment and a steady rate of growth. If the exchange is used to import I-goods to produce I-goods (i.e. for the build-up of heavy industry) then the capacity to produce I-goods will show a steady rate of growth, consequently the savings ratio will rise steadily and the economy as a whole will grow at an accelerating rate.

These possibilities are illustrated in the diagram below.

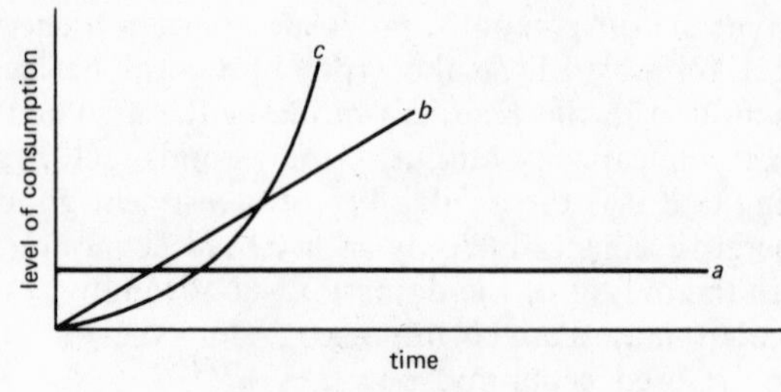


*a* = all *F* used for C-goods.
*b* = *F* used for I-goods to make C-goods.
*c* = *F* used for I-goods to produce I-goods.

FIG. 6.1

The Raj/Sen model is similar to the Feldman model except that the question at issue is the allocation of foreign exchange rather than the allocation of goods produced in the I-sector.

The assumption of a rigid foreign exchange constraint is the key assumption and also most subject to attack. Joshi (1970) and others[11] have shown the conditions necessary for an economy to be subject to a foreign exchange gap, distinct from a savings constraint. Joshi argues that there are two conditions necessary for 'a pure foreign exchange constraint': '(*a*) that the underlying rate of transformation in domestic production is zero over the relevant range and (*b*) that the rate of transformation through trade is zero, signifying that the elasticity of reciprocal demand is unity or less.'[12] Unless these (somewhat unlikely) conditions are met, then the foreign exchange constraint merges into a savings constraint.[13] Then additions to savings potential will be realisable in some increase in the savings and investment rate. However, while a pure foreign exchange constraint may be unlikely, with very low trade elasticities, the extra investment obtainable from a given reduction in consumption may be very small. In terms of the justification for a build-up of heavy industry, the pure Raj/Sen model requires a pure foreign exchange constraint. In the absence of such a constraint, then there is no special case for building up heavy industry as distinct from other industries. Very low elasticities, but not sufficiently low to create a pure foreign exchange constraint, may *in practice* justify the build-up of heavy industry for the same sort of reasons as those that lie behind the Raj/Sen model. *In theory*, in such a situation project evaluation should produce the required build-up, if appropriate shadow prices are applied, without giving any special weight to heavy industry. In practice, as an alternative to the continued application of shadow rates, the build-up in heavy industry might be encouraged directly (by subsidies or Government expenditure) in economies with very low trade elasticities and unrealised savings potential.

All the models which postulate that the investment rate depends on the capacity to make or import investment goods deny savings an independent role. It is assumed that in one way or another – by a shift to profits in a Keynesian (closed economy) distribution model, or by government tax and/or interest rate policy, investment rules the roost, and savings will not act as an independent constraint. But in LDCs savings may present an independent constraint, even if the Keynesian distribution model operates and/or government can tax as much as it likes.[14] This is because *consumption* is an essential input into the productive process, and cannot be depressed, below certain limits, without affecting the efficiency and even sometimes the possibility, of further investment. Real wages in the modern sector tend to be substantially higher (twice or more) than subsistence incomes outside. In part at least (see Chapter 3), the additional real wages, and consumption, are essential for efficient operation in the modern sector. As argued earlier, technology imported from advanced countries requires higher wages than the local subsistence sector.[15]

Whatever the direction and nature of causation – causation may well work both ways, with higher wages leading to more capital-intensive techniques and modern technology leading to higher wages – extra consumption is a requirement of extra employment in the modern sector. Suppose $w'$ represents the net extra consumption for each additional worker employed,[16] then

Necessary Consumption[17] $(C) = E \,.\, w'$, where $E$ is employment in the modern sector.

Hence the maximum savings possible is constrained by this requirement, or

$$S\,\text{max.} = Y - E \,.\, w'.$$

Capital goods output capacity (home-produced or imported) will only act as a constraint when

$$S\,\text{max} > I \,.\, \text{max}$$
$$Y - E \,.\, w' > Mt + F.$$

The level of necessary consumption is determined by the level of employment and the technology adopted.[18] The literature on choice of techniques considers variations in employment associated with a given amount of investment, and how, by appropriate choice of technique, $s/v$, or growth, may be maximised. In addition to necessary consumption, most economies also exhibit luxury consumption, that is consumption which can be curtailed without affecting productive efficiency. In practice, governments find it difficult to suppress luxury consumption, and thus in practice maximum savings may be lower than $Y - E \,.\, w'$.

In a capitalist economy, what appears to be luxury consumption may in fact be necessary for the efficient training and allocation of labour, and for the provision of adequate investment incentives. One of the most powerful determinants of investment in such economies is expectations about the rate of change of consumption. Thus one reason why consumption cannot be reduced to the minimum level necessary for an efficient work force may be the need to keep up consumption to ensure sufficient investment. In terms of the previous discussion, a high level of consumption is required to prevent the fourth constraint – willingness to invest – becoming operative.[19]

To summarise: growth may be limited by

(*a*) limitations on investment set by absorptive capacity;

(*b*) limitations on investment set by willingness to invest;

(*c*) limitations on savings and therefore investment set by difficulties in restraining luxury consumption;

(*d*) limitations on savings set by necessary consumption;

(*e*) limitations on investment set by the sum of local I-goods capacity and foreign exchange availability.

Special emphasis on build-up of the capital goods sector is only justified where (*e*) sets a lower limit than (*a*) to (*d*) above. Such a situation will only arise where the economy is closed, or where foreign exchange elasticities are very low, leading to a pure or near-pure foreign exchange constraint. The U.S.S.R. in the twenties and thirties probably came close to fulfilling these conditions: the economy was sufficiently advanced for absorptive capacity not to constrain investment opportunities, while willingness to invest did not arise as an independent factor. Luxury consumption could be ruled out, and since the economy was relatively advanced workers' consumption could be limited without affecting efficiency substantially. The economy was a near-closed one so that local investment capacity plus limited foreign exchange set the upper limit to investment. Hence, the build-up of heavy industry would seem to have been justified to speed up growth, and in the event (qualifying the model to allow for the war and heavy arms expenditure) the speed-up predicted did occur. In contrast, India in the 1950s and 1960s fulfilled few of the conditions. In particular, limited absorptive capacity reduced the returns from investment, though it did not impose a rigid bottleneck; luxury consumption proved unresponsive to government tax policy, partly, possibly, because it was necessary as an incentive to the working of the system. Above all, with a very poor economy, its savings capacity was limited by necessary consumption. While foreign exchange proved a continuous problem, it seems likely that this was, in large part, due to insufficient savings, rather than completely rigid foreign trade opportunities.[20] I-goods capacity as such therefore did not present for the most part the main bottleneck. This was shown, in the event, by the fact that, despite the build-up in heavy industry, the growth rate did not accelerate substantially, there was substantial spare capacity in heavy industry, and a continual savings problem.

The discussion has focused on Russia and India, but may easily be extended. Whether or not LDCs would be justified in building up heavy industry, according to the models discussed, depends on which of the limitations (*a*) to (*e*) operate and whether foreign exchange is a constraint. The extent and nature of the absorptive capacity limitation is likely to depend on the policies adopted towards investment allocation.[21] That is to say a policy which concentrates investment on one sector – in this case the capital goods sector – may come up against a shortage of well-thought-out viable projects sooner than a policy involving diversified investment in capital and consumer goods industries. This is especially likely in the short run so that any attempt suddenly to switch all investment to the capital goods industries would falter through lack of good projects, or what we have called the absorptive capacity limit, in many economies, before other bottlenecks became operative. However, in the longer run economies external to the firm but internal to the industry and managerial learning-by-doing within the capital goods industries may increase the absorptive capacity in that sector as a result of concentration of resources in the sector.

Few countries exhibit the relatively-high-income, closed-economy characteristics of the U.S.S.R. that would justify concentration on heavy industry. It seems likely that most LDCs will come up against absorptive capacity and savings constraints before rigid foreign exchange limitations, and special build-up of heavy industry would not therefore be justified. One might conclude, therefore, that the neo-classical position, discussed earlier, was close to being vindicated: such a position permits a premium on *savings*[22] without giving any special premium to the capital goods industry. However, if a premium on savings is combined with low foreign trade elasticities and a high shadow foreign exchange rate, in effect this will boost returns to projects in the capital goods industries, as compared with other projects.

## AN ALTERNATIVE APPROACH

All the models discussed have assumed away the problem of technology. All assume that the technology adopted is a given of the situation, a parameter and not a variable. The models may or may not[23] assume a choice of techniques, but the range of choice is itself one of the parameters of the system. But technology – that is, the nature of methods of production – is itself a product of the economic system. This has an important bearing on the question of the role of the capital goods industry.

Historically, the capital goods industries – particularly machine tools[24] – have been the prime developers of technology. In Britain, the U.S. and France, in the nineteenth century, machinery producers, rather than machinery users, led the way in innovation.[25] Freeman has shown that in the twentieth century the greatest concentration of research and development and innovation lies among heavy industries.[26] The central role of the capital goods industry in the development and diffusion of technical change is due to a number of factors: first, and most obvious, new products and new processes generally require new machines.[27] Thus without capital goods production many new ideas will remain on the drawing-board. But the capital goods sector does more than simply enable ideas to be realised: it is also a major initiator of change. One reason for this is that a major source of market expansion for machinery producers lies in the replacement of existing machines; development of new machines, and hence rapid obsolescence of old machines, is a powerful instrument for securing rapid replacement. In addition, there is considerable technological feedback within the machine-making sector, with developments at one stage stimulating and often requiring developments elsewhere. An innovation may induce subsequent innovations in consuming or supplying industries because of the changed scale of requirements following the innovation, or because of changed technical requirements. Strassman has emphasised the scale factor; Rosenberg, technological imbalance. Innovations in the machine sector are most likely to stimulate themselves because they are both input and output, part of the circular process of production.[28] In contrast, innovations in consumer goods – since they do not form an input into further production – are likely to stimulate fewer further innovations. Strassman has stressed that the greater interrelatedness of capital goods production is likely to make an innovation in this sector more productive of innovation in the economy as a whole than innovations elsewhere. The net result is that innovation is likely to be concentrated in the capital goods sector – as observed; and that economies without a capital goods sector are more than proportionately weakened when it comes to innovatory activity. Their innovations tend to be dead-end innovations, rather than a cumulative process. Put in another way, there are greater externalities to innovation in the capital goods sector than innovation elsewhere, and the capital goods sector is likely to benefit more than other parts of industry from externalities caused by innovations elsewhere.

Historically, the capital goods sector has also diffused innovation, spreading new ideas across sectors as well as within them. There is also an important learning and training dimension to capital goods production, particularly in relation to innovation. Countries which have no capital goods sector tend also to lack 'the base of skills, knowledge, facilities and organisation upon which further technical progress so largely depends'.[29]

Historical and analytic approaches both conclude that a capital goods sector is essential for innovatory activity. Lacking such sectors, underdeveloped countries have to import not only their machinery, but also their technical progress. The nature and direction of technical progress is thus determined from the outside.

Capital goods are part of the technology package. Associated with a particular capital good is a set of requirements that go with it, which are difficult, often impossible, to separate from it. The nature of different types of technology package has been discussed at much greater length in Chapters 1, 3 and 4. The characteristics associated with a particular capital good include, for example, the nature and specification of the product, requirements for inputs, skill and managerial requirements, wage levels, etc. In the absence of capital goods industries, countries are forced to import capital goods from the capital goods producers – chiefly the advanced countries – and therefore to import from them the whole technology package. As argued in Chapter 3 this package, designed for different circumstances, is largely inappropriate to the circumstances of underdeveloped countries, and its import causes inappropriate and undesirable patterns of development.

To get technological change responsive to the conditions in the LDCs the change must originate in the LDCs. For this capital goods capacity is essential to build up skills associated with technological change, and to realise appropriate technological change. As a first step a capital goods industry may produce capital goods previously produced in developed countries, but no longer produced because obsolete there.[30] But using more recently developed techniques and materials, old designs may be modified, increasing their efficiency and developing characteristics more appropriate to the conditions of the countries in question. The existence of a capital goods industry is likely to stimulate technical change, while the fact that the industry is serving an underdeveloped country is likely to make the technical change responsive to the needs of that country.

The existence of a capital goods industry is, of course, not a sufficient condition for appropriate technical change, though it is a necessary one. The Indian capital goods industry has for the most part imported advanced technology; it has neither adopted (on a significant scale) old designs, nor has it adapted new ones.[31] One explanation of this may be the philosophy behind the development of the industry, which was the Mahalanobis one of a rapid build-up in capacity, rather than any idea of generating appropriate technical change. To secure this rapid build-up required the use of extant foreign technology. The emphasis was on expansion of the industries which contributed most quantitatively to

I-goods capacity, e.g. the iron and steel industry, rather than on those industries most likely to generate technical change. For rapid quantitative expansion of I-goods capacity, large-scale firms using modern Western technology and employing Western technicians may be necessary.

The type of capital goods industries likely to lead to technical change is very different. Small-scale, locally financed firms, with limited access to foreign sources of technology catering for firms similarly placed, are most likely to be productive. While a Mahalanobis-type approach suggests concentration on the big input sectors, like iron and steel, and chemicals, technical change is more likely to come from machine-tool makers. One difficulty facing underdeveloped countries is that a specialised capital goods manufacture is more likely to innovate, but specialisation requires scale, and most countries' markets are too small to provide the required scale. Scale in the market for capital goods might be attained through specialisation – and hence scale – in the production of final goods. Trade between underdeveloped countries in capital goods would also help.

## CONCLUSION

The first part of this chapter showed that the build-up of capital goods industries in LDCs on the basis of a Feldman type analysis is rarely justified, because the assumptions behind the model rarely apply. However, to conclude that there is therefore no justification for giving special encouragement to capital goods industries in LDCs, beyond that suggested by the immediate returns on the project, is rejected in the second part of the chapter on the grounds of technological development. It is argued that a capital goods sector is an essential condition for local technological development, and that without such development LDCs are forced to accept the technical change of the advanced countries, with deleterious consequences for the rate and pattern of development. Build-up of capital goods industries is therefore justified on technological grounds. However, this leads to a different type of capital goods industry than the sort of industry which would be justified by the earlier Mahalanobis-type models.

## NOTES

1. For the Soviet Union see Erlich (1960) and Preobrazhensky (1926); and for India see Mahalanobis (1953), Raj (1961), Bhagwati and Chakravarty (1969), Bhagwati and Desai (1970).
2. Marx includes raw materials production along with capital goods in Department I.
3. Popularised in the West by Domar (1957). Hans Singer (1952) put forward a similar model. Raj and Sen (1961) further developed the Mahalanobis categorisation to allow for different types of investment goods, for intermediate goods and for a limited amount of international trade.
4. 'The capital goods in existence at a moment of time are all the goods in existence at that moment.' (Robinson, 1953–4)

5. Marx, *Capital*, Vol. 2, Chapter XX, Section II.
6. Different economists have picked out different aspects. For example much business cycle theory has been concerned with the relationship between investment capacity and willingness to invest, while Keynes' prime concern was the relationship between savings capacity and willingness to invest.
7. This presentation is derived largely from Domar, but to simplify it, it is assumed that the ICOR is the same in both sectors.
8. For some empirical evidence see Wilber (1969) on Russia, and Bhagwati and Desai (1970) on India.
9. The Feldman model is slightly odd in this respect since it assumes that once they are built there is no substitutability between machines for production in the I-sector and machines for production in the C-sector, yet it also assumes that within the I-sector output may be costlessly switched to producing machines for the I-sector. There may well be rigidities and lack of substitutability which limit this kind of switch, and hence limit the freedom to select $\lambda$, the proportion of output going to the I-sector.
10. However, there may well be discontinuities, so that decreasing returns are, as investment increases, suddenly reversed and replaced by increasing returns, as some key investment (e.g. in energy) takes effect.
11. See, e.g., Lal (1970).
12. Joshi (1970), p. 115.
13. Again these criticisms can be seen as part of the general attack on any bottleneck model.
14. For fairly obvious reasons, it is unlikely that this 'even if' clause does apply in most economies.
15. Kidron (1974) discusses the difference between wages in the advanced and underdeveloped countries in these terms, arguing that the higher nominal wage-rates in advanced countries are necessary, given differences in technology and worker efficiency, and not due to a greater 'luxury' element in advanced-country wages.
16. To calculate the additional consumption generated by shifting a worker to the modern sector one must take into account his extra consumption, and the effect on consumption on those outside the sector, while the net effect on savings also depends on the change in production outside the modern sector. It is assumed here that $w'$ is calculated taking these factors into account. For a formula for calculation of the effect on savings and consumption see Little and Mirrlees (1969), Chapter XIII.
17. Marx's distinction between 'consumer *necessities*' and 'articles of luxury' (*Capital*, Vol. II, Ch. XX, section IV) is much the same as that adopted here, but he did not include what might be described as 'incentive' consumption as being necessary for the workings of the capitalist system.
18. Given a positive relationship between the level of necessary consumption per worker and the technology used, as argued above, the wage-rate is likely to vary with the choice of technique, being higher in more sophisticated and investment-intensive techniques. This means that the savings potential from more investment-intensive techniques is relatively lower than normally assumed in the literature (see e.g. Dobb and Sen), where it is assumed that the wage is invariant with respect to technique chosen.
19. In theory, in open economies exports should be able to provide the required incentives – this is the view adopted by those who believe in 'export-led growth'. In practice it seems that export markets depend on prior investment, and hence cannot, initially at least, provide the incentives for such investment.
20. See Bhagwati and Desai (1970), and Nayyar (1976).
21. I owe this point to Charles Kennedy.
22. Under special assumptions discussed in Little and Mirrlees (1969) and the UNIDO Guidelines by Das Gupta *et al.* (1972).

23. Contrast Domar (1951), who assumes a single technique for each sector, and Sen (1968), who is primarily concerned with choice of techniques.
24. See e.g. Rosenberg (1963(b)) and Habakkuk (1962).
25. See e.g. Landes (1969), Rosenberg (1963(b) and 1969) and Saul (1967).
26. 'The industrial pattern of research expenditure is strikingly similar in Britain and America. In both countries one group of industries – mainly capital goods and chemicals – account for over nine-tenths of research expenditure.' Peck found that in the aluminium industry, while product innovations originated largely with primary producers, process innovations were initiated by equipment makers rather than end-product users or primary producers.
27. See Freeman (1968) and Rosenberg (1963(a)).
28. Many of the most famous developments of the industrial revolution were due to technological imbalances as one technical development speeded up one aspect of a process, creating a new bottleneck, which required further innovation to break: the famous innovations in textiles provide examples. For other detailed examples, see Rosenberg (1963(b)).
29. Rosenberg (1963(a)).
30. See Pack and Todaro, who see this as being the essential function of capital goods industries in LDCs.
31. For evidence of this on the steel industry in India, see Johnson (1966); also Leff (1968). A survey of the engineering industry in Colombia concluded 'that much of the equipment and many of the designs are imported from abroad'. But some modifications to imported machines and designs were also noted. See the Report of the Institute of Technological Research in Bhalla (ed.) (1975).

# 7 Trade and Technology

The discussion so far has paid little attention to international trade. Yet a country's trading relationships heavily influence patterns of production and consumption, and therefore the technology in use. This chapter considers the relationship between trade, development and technology.

Conventional theories of international trade take technology as a parameter of the system, just as it is taken as a parameter in production theory. One of the many assumptions behind Heckscher-Ohlin[1] (H-O)-type analysis is that tastes and technology are given and unchanging. It is assumed that all countries have equal access to the technology – i.e. have equal knowledge – and face the same potential production function. They differ in their resource availability, or factor endowment, and the gains from specialisation and international trade arise from these differences. Broadly, the rich countries (referred to for brevity in what follows as 'the North') are capital-abundant and therefore specialise in capital-intensive goods and processes, which they exchange for labour-intensive goods produced in the labour-abundant poor countries (the South).[2] Both North and South gain by the exchange – both moving on to a higher social welfare curve than is possible in a no-trade world, as shown in Figure 7.1.

In so far as differences in comparative advantage arise from differences in factor availability, there is likely to be most international trade, and the gains will be greatest, between countries which are *dissimilar* in terms of factor endowment, and least between countries which are *similar*: so that trade should be greatest between North and South, and least within each group (South/South and North/North). According to these assumptions, if countries have identical factor availability they will also have identical production possibilities (since they have equal access to technology); and in the absence of economies of scale and learning-by-doing effects, as assumed by the theory, there would be no scope for trade. Differences in the availability of natural factors – land, climatic conditions and mineral resources – also account for trade flows. Trade in commodities for which location of production is determined by natural conditions[3] has been described as 'Ricardo goods' trade.[4]

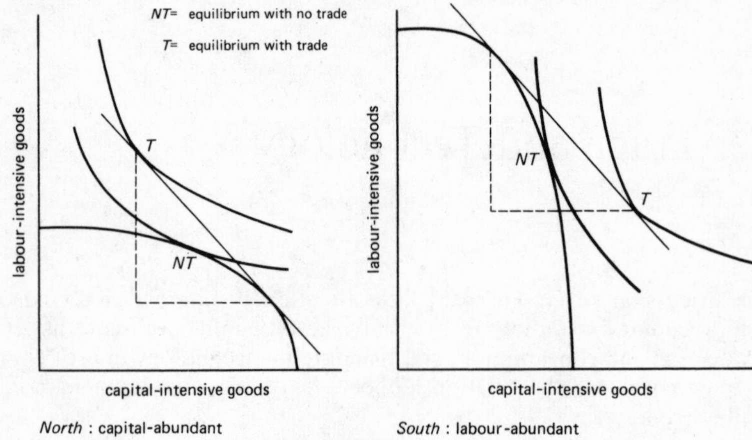


FIG. 7.1

Thus the Heckscher-Ohlin-type analysis would predict flows of trade as depicted in Figure 7.2, with capital-intensive goods moving from North to South, labour-intensive moving from South to North, and Ricardo goods moving from where natural conditions locate their production to the consuming centres, which means for the most part from South to North. Not only do both areas gain from trade, but ultimately, given a particular set of assumptions, factor-price equalisation occurs[5] – that is to say the export of labour-intensive goods has the same effect on the South as the export of labour would, diminishing its relative abundance, and therefore increasing its price, until ultimately the relative scarcity of labour and capital is the same in North and South.

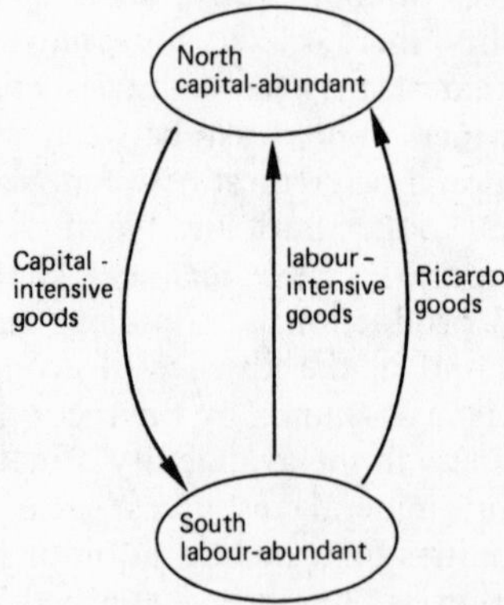


FIG. 7.2

The predictions of Heckscher-Ohlin and Stolper-Samuelson have been shown to be factually incorrect in three respects.

TABLE 7.1 Proportion of total exports, 1973 (manufactures in brackets)

| *From* | *To North* | *To South* | *To centrally planned* |
|---|---|---|---|
| World | 71·3 (69·4) | 18·2 (19·7) | 9·7 (10·5) |
| North | 76·7 (75·4) | 18·1 (19·8) | 4·5 (4·4) |
| South | 74·9 (67·8) | 19·7 (28·3) | 4·4 (3·3) |
| Centrally planned | 26·9 (16·9) | 15·4 (12·8) | 56·7 (69·6) |

Proportion of total imports, 1973

| *To* | *From North* | *From South* | *From centrally planned* |
|---|---|---|---|
| North | 76·3 (88·6) | 19·9 (8·6) | 3·8 (2·8) |
| South | 70·9 (82·6) | 20·6 (11·2) | 8·5 (6·2) |
| Centrally planned | 33·1 (37·4) | 8·6 (2·8) | 58·3 (59·8) |

Change in value of world trade (% p.a.), 1960–73

| | *Total* | *Manufactures* |
|---|---|---|
| North–North | 12·7 | 15·1 |
| North–South | 9·8 | 10·3 |
| South–South | 10·1 | 14·7 |
| South–North | 11·5 | 20·5 |

NOTE North/South groupings exclude centrally planned economies.

SOURCE *U.N. Monthly Bulletin of Statistics*, December 1971, July 1975.

*One*: trade flows are greater between rich countries and regions (North/North) than between rich and poor countries (North/South).

As Table 7.1 shows, the developed countries account for about three-quarters of total trade and a greater proportion of manufactures. There is no tendency for trade between countries with dissimilar factor endowments (North/South) to be greater than trade between similar countries. In fact the proportion of trade accounted for by North and

South countries is similar for both North and South, with a definite tendency for greater trade between *similar* groups than dissimilar groups. This tendency is more marked in trade in manufactures, where the Heckscher-Ohlin factor proportions theory ought to apply most, than in trade in primary products where natural factors are often of over-riding importance in determining the direction of trade.

*Two*: Leontief, in a famous 'paradox', showed that U.S. exports were labour-intensive relatively to her imports. This paradox has been 'explained' by the heavy human-capital content of U.S. exports. This explanation, which takes us far from the simple factor endowment view, tends to turn the theory into a tautology.[6] There is a further difficulty about using *factor endowment* as an independent explanation of trade. In its use to describe natural resources – e.g. climate, and therefore agricultural product potential, mineral resources availability – endowment may be taken as largely independent of economic development and trade. But once one uses it to describe man-made factors, like capital goods, educated manpower, etc., then the endowment is itself a product of past trade developments; current policies cannot therefore take factor endowment as given but must regard factor endowment as one of the variables which may be affected by policy.

*Three*: the Samuelson/Stolper development of H-O also obviously lacks factual support. Far from equalisation of factor prices, divergencies between real wages in rich and poor countries seem to have been increasing, absolutely though possibly not relatively.[7] All agree that the highly restrictive assumptions of the Samuelson/Stolper theory have not been met, and therefore the theory has not been tested. The interesting question, however, is whether the particular assumptions that have not been met are of a policy nature – i.e. the imposition of trading restrictions – or derive from the fact that the nature of the process of economic development differs from their assumptions. In the former case, freeing trade might contribute to factor price equalisation; in the latter, it might well accentuate differences.

Many of the required assumptions[8] of the Heckscher-Ohlin theory are far from realistic. Among the necessary assumptions are the absence of economies of scale, the presence of perfect competition (nationally and internationally), full employment of resources in both countries, identity (or very close similarity) of production functions (and therefore also of technical knowledge) and patterns of tastes in all countries concerned, and the absence of tariffs and other barriers to trade. In addition, the theory is *static* with the parameters – tastes and technology – assumed to be unchanging over time.

The obvious invalidity of many (possibly all) of these assumptions does not necessarily invalidate the approach. As Johnson summarises the position:

> These assumptions are, like the assumptions of all theorising, abstractions from the complexity of reality capable of refutation as universally valid assumptions by an appeal to the facts of observation: the scientific issue however is whether observed deviations of facts from assumptions are empirically significant enough to destroy the validity of the conclusion of the theory.[9]

Relaxation of some of the assumptions can be fairly readily incorporated into the theory: thus the presence of monopolistic conditions in international trade has led to the justification, from a purely national point of view, of an optimal tariff. But others are less easily dealt with: for example, the absence of full employment of resources can destroy conclusions based on the opposite assumptions, as a movement towards the fuller use of resources may outweigh the effects on allocational efficiency. But here we are mainly concerned with the consequences of abandoning the restrictive assumptions about technology. These assumptions, to repeat, are that countries face near-identical production functions, that these production functions exhibit constant returns to scale and diminishing returns to the increased use of a single factor, and that they are unchanging over time. In addition tastes, which are also part of the technology package, are assumed given and unchanging.[10]

Alternative approaches to international trade which question these assumptions may be classified into four strands:

- the Myrdal (and more recently Griffin) argument about cumulative causation;
- the Burenstam Linder hypothesis on the relationship between income levels, consumption patterns and trade;
- the product cycle argument associated with Vernon, Hirsch and others;
- the appropriate products/technology argument.

We shall deal with these in turn.

## I INCREASING RETURNS AND CUMULATIVE CAUSATION

Myrdal's basic theory of cumulative (and circular) causation is very simple: it is that any initial advantage in cost which leads to an expansion of output will generate further cost (and other) advantages which will reinforce the initial advantage and lead to cumulative growth; conversely an initial disadvantage leading to a reduction in output will generate cumulative diseconomies and contraction. Thus free trade – within a country or between countries – far from producing income equalisation will tend to lead to wide and increasing disparities in growth and levels of income.

Cumulative causation arises from three types of increasing return:

(i) static and internal to the firm;
(ii) static and external to the firm;
(iii) dynamic.

Increasing returns[11] mean that a country's so-called comparative advantage depends not only on factor endowment, but also on size of market. Countries with large home markets (developed countries) have an initial cost advantage, irrespective of factor endowment. If increasing returns are limited to a certain quantity (the flat *U* cost curve or inverted *J* curve), then initial protection (combined possibly with regional trade agreements) may enable developing countries to reach the critical size. With indefinitely increasing returns, the larger market is *always* at a cost advantage. Only if, by a huge subsidy, the developing countries can capture a larger market than the developed countries, will they be able to compete; and this will involve heavy (and risky) subsidisation of one industry, not the across-the-board protection carried out by most import-substituting countries.

If increasing returns are external to the firm but depend on the total size of the industrial base, then subsidising particular products will not be enough; general industrial expansion is required. Dynamic increasing returns mean that, quite apart from static economies of scale, over time productivity is increased faster, the greater the size of the industrial base – i.e. the rate of growth is positively related to the level of output (and to past changes in output).[12] This may be partly due to the fact that resource accumulation – foreign and domestic savings, migration of skilled and unskilled labour[13] – is positively related to the level and change in output. It is also due to the fact that technical progress, or the increase in output obtainable from any increase in inputs, is likely to be higher, the higher past levels and growth of output. This arises from concentration of R and D expenditure on growth points, and learning by doing.

The evidence for some sort of cumulative causation is considerable,[14] and certainly far outweighs any evidence for neo-classical factor price equalisation. There are, however, some breaks on the process, as shown by the varying industrial experience of different countries, and regions within countries, over time. One such brake arises from restrictions (natural and artificial) on the free movement of factors, particularly unskilled labour; another may be a stultification of social structure and attitudes which eventually slows down the growth of mature economies. However, the forces making for cumulative causation, particularly at the relative stage of development of underdeveloped and advanced countries, are operative in most cases. In such a situation free trade is liable to lead to specialisation along lines of *comparative disadvantage*, from a growth point of view. North/South trade is particularly liable to lead to cumulative growth for the advanced countries and static development for the underdeveloped countries. In contrast South/South trade does not contain the same dangers, though even here protective devices are required to prevent some (the more developed) parts of the South from benefiting cumulatively, and others losing. This is one of the major problems for any proposal for increased integration among underdeveloped countries.

The theory of cumulative causation does not necessarily disagree with the Heckscher-Ohlin predictions about the direction of trade (see Figure 7.2 above) but emphasises that the dynamic implications are likely to be increasing relative growth for the North.

## II SIMILARITY OF STAGE OF DEVELOPMENT

Burenstam Linder argues, in direct contrast to the neo-classical theory, that trade flows are likely to be greatest between countries at a similar stage of development. This arises from the close relationship between income level, consumption and production patterns. Consumption patterns, it is argued, depend on income levels, with different types of product consumed at different income levels. In this respect the argument is very similar to that about the relationship between products and income level developed in earlier chapters. Countries which are at a similar stage of development, therefore, consume similar goods. Innovation and production initially take place in order to supply the home market. Once a product is established on the home market, the industry widens its horizons to markets in countries with similar income levels, since the product was designed for that type of consumption pattern.

The Burenstam Linder predictions about the directions of trade, summarised in Figure 7.3, are in stark conflict with the H-O predictions. The difference arises because for Burenstam Linder production for consumption in the home market precedes exporting, which means that any goods which are exported are first designed for home consumption. Export markets arise in similar economies. The H-O theory divorces production from consumption, so that a country produces whatever it has a comparative advantage in, irrespective of whether the good is consumed domestically. The latter view is reasonable in a world of unchanging technology, perfect knowledge and perfect markets. In such a world there is no need to seek the security of the home market for new processes and products before venturing on to the world market. But in the real world of innovation and uncertainty, where firms are forging production possibilities rather than choosing between them, the Burenstam Linder view fits in with what we know about innovation and trade in manufactured goods.[15]

It thus offers an explanation of the very large weight of North/North trade in world trading patterns, contrary to H-O predictions. But the

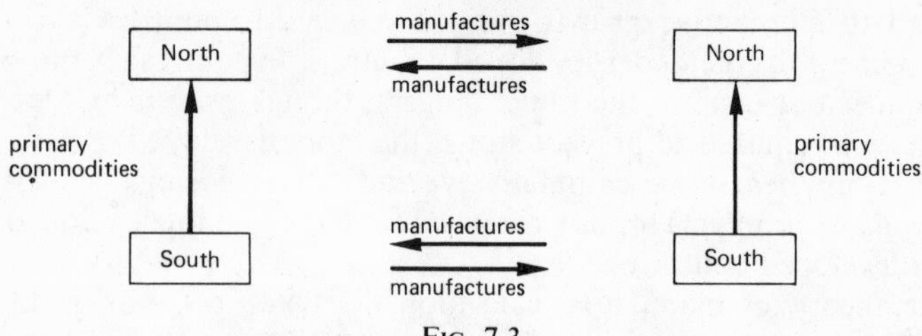


Fig. 7.3

thesis would suggest that South/South trade should similarly predominate among South countries' trading relationships – yet as Table 7.1 shows only one-fifth of total South countries' trade goes to other South countries, though the proportion is higher (28 per cent) for manufactures. One reason for the predominance in South trade of trade with the North, which accounts for 75 per cent of total South exports and 68 per cent of manufactures, is historical relationships, colonial and neo-colonial. But another part of the explanation lies in the concentration of technological innovations in rich countries, and the consequences of this concentration for patterns of trade. The last two approaches are concerned with different aspects of this concentration.

### III PRODUCT CYCLE

Technology developed in one economy may be transferred internationally. Consequently, while it is normally true, as Burenstam Linder argues, that innovation and development of a product first takes place in the country for which there is home demand for the product, subsequent location of production may be transferred by the transfer of technology, either via the multinational or by sales of the technology. Thus an initial innovation may lead to trade between the innovating country and other countries of similar demand structure (technological gap trade), but subsequently the technology may be transferred and the trade eliminated.[16]

However, so long as innovation occurs continuously, new innovations create new technological gap trade, so that as previous gaps close through the international transfer of technology, new gaps and new technological gap trade opens up. The theory of technological leads to explain trade was initially put forward to explain trade patterns between developed countries, by Posner.[17] All the countries could be assumed to be technological innovators. Consequently, one country might have a technological lead in one good and hence net exports of that good, while the gap lasted, and other countries would have leads, and net exports, in other goods. According to these views, technological innovation and technological gap trade explained trade between similar countries, on much the same lines as the Burenstam Linder explanation, except that the technological lead in any good, and the resulting trade, was assumed to be temporary because after a while the technology would be transferred (or developed), eliminating the technological lead and the trade. The explanation of trade was designed primarily to explain trade between innovating countries – North/North trade. South/South trade of the same type would not arise because of the absence of innovation in the South. North/South technological gap trade would arise, but it would be all one way – North to South – because of the concentration of innovation in the North.

The product cycle theory proper[18] systematised the relationship between North and South via innovation in the North and subsequent transfer of technology to the South. Innovation and initial production takes place in the developed countries. But subsequently, when the product (and associated technology) is debugged, and markets are well established, multinational firms may transfer production to cheap labour areas (the South); from there the product is exported back to the North for consumption there. The possibility of transferring production location in this way destroys the neat (Burenstam Linder) relationship between exports and stage of development, since though initially innovation takes place in the home market, subsequently production location may be transferred to quite dissimilar countries – indeed the point of transferring production location is to benefit from the different factor availability and price in other countries and therefore production is likely to be transferred to dissimilar countries. Thus South/North flows of goods are created as the product cycle goods return to the market for which they were designed. A variant of this occurs when multinationals locate labour-intensive parts of the productive process in the South to benefit from the cheap labour, and re-export the processed output back to the North.[19]

North/South trade in manufactured goods can then be seen mainly as a form of technological gap trade, consisting of goods innovated in the North before the technology is transferred to the South. The Burenstam Linder thesis might be taken to imply that such trade should not occur, because given the dissimilar stage of production and per capita income levels between North and South, there should also be dissimilar consumption patterns, with a different set of goods consumed in the two parts of the world. But this overlooks two factors: *first*, the concentration of innovation and marketing skills in the North to such an extent that South products *hardly exist*. Because of the concentration of scientific knowledge and expenditure on innovating in the North, their products are for the most part substantially more efficient than South products, fulfilling a greater variety of needs and doing so more efficiently in relation to resource cost.[20] Hence, in large part, the North products meet no competition from South products, which would otherwise, and *ceteris paribus*,[21] be more appropriate to the South stage of development and per capita income. *Secondly*, markets depend not only on the level of per capita income, but also on its distribution. If unequally distributed, consumers with similar income levels and patterns of consumption may be thrown up, even though the overall per capita income levels are quite dissimilar. This, in effect, is what has happened in many poor countries, where a market for the products of advanced countries has developed among the elite.[22] As argued at greater length in Chapter 3, the inequality of incomes in many underdeveloped countries is itself associated, as consequence as well as cause, with the consumption of high-income (North-developed) products, and with the production of such products using technology developed in the North. Production using North techniques tends to lead to some inequality as the production patterns of the North, including (to some extent) patterns of factor payment, are replicated. Apart from the consequence of using developed-country technology on factor payments and the distribution of income, the *consumption* of high-income products tends to reinforce inequality. This occurs most directly in the case of public goods, where the provision of North goods concentrates public expenditure on a small section of the population. In the case of private goods, the effect is more indirect and is related to government policy on income distribution.[23]

The thesis that trade in manufactures occurs mainly between similar countries is thus correct in explaining North/North flows of goods. But concentration of innovation in the North explains North/South and South/North flows of manufactured goods. The North/South trade is a form of technological gap trade, encouraged by unequal income distribution in the South, plus the complex of factors that are described as the 'demonstration effect',[24] which create a market in the South for high-income North products, particularly in the absence of efficient South products. South/North trade in manufactures consists (mainly) of the export of goods originally developed in the North, whose technology has been transferred to the South to take advantage of the lower labour costs. Comparative advantage enters the picture in that it determines which type of technology and production it is worth transferring; thus products (or parts of products) which use labour-intensive technology are more often transferred than skill and investment-intensive goods. Thus the empirical findings that very broadly trade flows are in accordance with comparative advantage[25] in nature (though not in magnitude, as illustrated in Table 7.1) are due to the selective transfer of technology, in a dynamic process of innovation and transfer, rather than the static view of specialisation along a given production frontier, as presented by Heckscher-Ohlin etc.

Trade restrictions change the pattern of technology transfer. Restrictions on the entry of labour-intensive goods into the developed countries discourage technology transfer which involve re-export of goods back to the North, while the across-the-board restrictions in underdeveloped countries encourage similar across-the-board technology transfer, irrespective of factor use.

The trade flows, according to this view, are summarised in the diagram below:

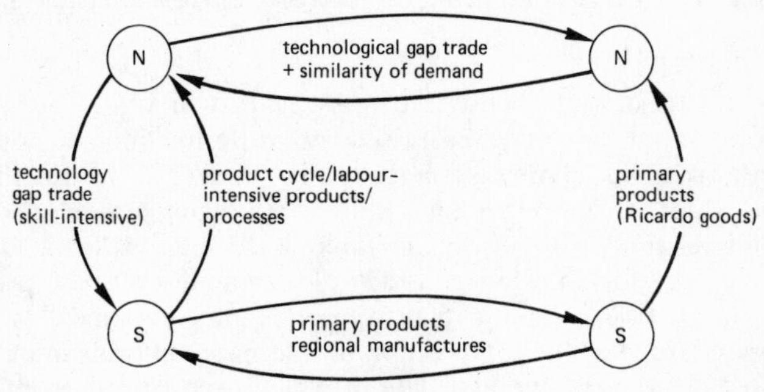


FIG. 7.4

TABLE 7.2
Proportion of total exports (1974) to developing market economies going to:

| | *Developing Africa* | *Middle East* | *Other Asia* | *Developing America* |
|---|---|---|---|---|
| *From* | | | | |
| Developing Africa | 34·7 | 5·8 | 9·7 | 48·3 |
| Middle East | 7·5 | 19·0 | 39·4 | 33·6 |
| Other Asia | 7·4 | 11·7 | 71·4 | 8·2 |
| Developing America | 4·7 | 2·5 | 3·2 | 89·5 |

Proportion of manufacturing exports (1973: S.I.T.C. sections 5, 6, 7, 8 excl. 68) to developing market economies going to:

| | *Developing Africa* | *Middle East* | *Other Asia* | *Developing America* |
|---|---|---|---|---|
| *From* | | | | |
| Developing Africa | 68·0 | 11·4 | 11·7 | 9·3 |
| Middle East | 15·4 | 76·7 | 7·1 | 0·4 |
| Other Asia | 10·4 | 9·8 | 72·1 | 5·9 |
| Developing America | 2·6 | 1·5 | 3·8 | 91·9 |

SOURCE U.N., *Yearbook of International Trade Statistics*, Vol. 1, 1974, Special Table B.
U.N., *Monthly Bulletin of Statistics*, December 1975, Special Table B.

There is proportionately little trade in manufactures between South countries – since in the absence of innovation in the South there are no South products to allow either technological gap trade, or similarity of demand trade. What trade there is tends to be concentrated regionally (as Table 7.2 shows), often being the outcome of regional trade arrangements.

## IV APPROPRIATE TECHNOLOGY AND APPROPRIATE TRADE

Technology changes continuously and with it production and consumption patterns. This continual change has its parallel in the flux of international trade. Bruno (1970) has shown how comparative advantage changes over time, with any league table of the relative factor intensity of different goods showing considerable instability as new processes transform previously labour-intensive goods into investment-intensive. Hufbauer's study of the composition of international trade in manufactures in 1965 showed the first date at which each traded product had been included as a special category in the Standard Industrial Trade Classification. No country had an *average* first date earlier than 1944. That is to say, typically, the composition of traded products had been entirely changed after seventeen years. This approach only included changes big enough to justify a new S.I.T.C. category. Many other process/product changes occur within the same category. The technological gap/product cycle approaches to international trade emphasise the continuously changing character of international trade, following the changes in technology: innovations create exporting opportunities; technology transfer obliterates them, sometimes reversing the flow; while new innovations create new flows.

The essentially dynamic and changing character of the flows has important implications for trade strategies for the third world. First, it means that a trading strategy demands continual adjustment,[26] requiring further technological transfer involving continued technological payments and dependence, heavy managerial costs,[27] and much uncertainty – in itself enough to throw considerable doubt on the strategy. Secondly, the changes have consequences for the technology adopted. A strategy involving free (or near-free)[28] trade between North and South would appear to allow the South to specialise on and export labour intensive products, in exchange for skill and/or capital-intensive goods from the North. Compared with an autarkic import-substitution policy it would appear to offer a more employment-using technology, as well as the more conventional gains from trade. Put in another way, it should permit the fuller use of factors, or a movement towards the production frontier, as well as gains from comparative advantage, or from movement along the production possibility curve.

The gains from alternative strategies depend on the precise nature of the alternatives. The comparison between export orientation and import-substitution is critically dependent on the nature of the import-substitution policy. Policies of *import reproduction*,[29] involving the exact replication of goods previously imported from the rich countries, require the transfer of the technology used to produce the goods in the advanced countries and its replication in poor countries. This means the transfer of techniques developed and used in the advanced countries, generally inappropriate to the poor countries. In addition to the inappropriate factor requirements – investment- and skill-intensive – across-the-board replication means that each product is produced on a far smaller scale than that for which the technology was designed. The result is low-capacity utilisation, high-cost production, and minimal employment creation[30] – indeed all the consequences of importing inappropriate technology discussed in Chapter 3. An export strategy has to be *feasible*: it is not possible before some sort of industrial base has been established, which is why Ranis and Fei have argued that an initial import-substitution policy is necessary for subsequent export success. It also requires that potential importing countries permit the entry of the exports. This condition is more likely to be met for small countries than large, and for countries with particular ties with the potential importers. But given feasibility, an export strategy may, in the short run, permit gains as compared with an import-substitution policy. But the initial gains of such a strategy may rapidly be offset by the dynamic adjustments required.

To maintain a share of international markets, continual product improvements are needed. So long as innovation continues to be monopolised in the North, this requires continuous import of Northern technology, and corresponding adjustments to processes and products. But the later vintages of such technology will be more skill- and investment-intensive, larger-scale, in tune with the changes in the advanced countries, as described in Chapter 3. Consequently, a strategy of trade with the North will require the adoption of continuously less labour-using technology over time.

A strategy of trade with the North requires continual interlocking with North technology. On the import side, such a strategy involves a changing composition of imports in line with the changing composition of international trade, which means changes in line with changes in consumer patterns in advanced countries. The proportion in which different types of goods are imported – and to a greater extent the proportion in which different types of goods are consumed – may differ from the proportions obtaining in North countries. But the products themselves – when imported from the North – must accord with what is being produced in the North. The changes in the characteristics of goods that occur with technical change over time, reflecting greater technical possibilities and higher income levels, will therefore also apply to the imported goods consumed in the poor countries. This is liable to mean that while the goods become more efficient in need-fulfilment over time, they also have higher-income characteristics, becoming increasingly inappropriate, as argued in Chapter 3. Imbalance in consumption patterns and *increasing* inequality will be needed to provide a market for such goods. Thus, while on the export side trade with the North tends to require product and process changes involving increasingly inappropriate patterns of production, on the import side a strategy allowing the free import of goods from the advanced countries means that the consumption pattern becomes increasingly inappropriate. Hence the short-run gains of such a strategy, in terms of employment, may be offset in the long run by effects on employment, consumption and income distribution.

It may be suggested that this argument depends on the assumption that innovation is monopolised in the North. If South innovation occurred, then the strategy could be successfully pursued without the deleterious consequences. But this is an illusion for two reasons: first, successful innovation might enable exports to keep up with world competition, while using locally produced and more appropriate technology. But this would leave the import side untouched. Concentration on exports would lead to innovation in techniques for use locally to produce goods to be consumed in the advanced countries. Hence the products would need to be designed for consumption in high-income markets, and therefore no innovative effort would be made to develop appropriate low-income products. Appropriate techniques might ensue but not appropriate products for the home market. Innovation in the North is unlikely to be directed towards low-income products because innovation is generally directed first at the home market, and subsequently spills over into exports,[31] and because the third world provides only a small proportion of the markets of the advanced countries. Therefore, the later products, whether produced locally or imported, would tend to be increasingly inappropriate. The second reason for doubting the strategy of trade with the North plus South innovation lies in the nature of innovation. It is unlikely that South innovation could be successfully pursued in the context of a trade with the North strategy: this is partly because innovation is an infant industry which requires initial protection. The failure (in large part) of Indian innovation[32] even on the Indian market is due to the lack of protection, as compared with imported technology. Producers dare not risk local technology, given consumer preferences for foreign products and brand names. While this has prevented local innovations being adopted on any scale within third world countries for home consumption, it would be a much stronger factor on world markets, particularly since it would be reinforced by all the restrictive arrangements which maintain world markets for the established multinationals.[33] An export-orientated strategy is unlikely to be consistent with successful local innovation.

While an export strategy towards North markets is likely to be combined with increasingly inappropriate technology, the same may well be true of an import substitution policy, if the policy involves changing patterns of import substitution in line with changing patterns of production and consumption in advanced countries. This, broadly, has been the strategy adopted in many Latin American countries, as described by Furtado:

> The industrial nucleus linked with the domestic market develops through a process of displacing importation of manufactured goods. . . . The greatest concern of the local industrialist is therefore to provide an article similar to the one imported. Thus the technological innovations which appear most advantageous are those making it possible to approach the cost and price structure of the developed countries.[34]

This type of import-substitution strategy requires continuous adjustments in line with changes in world technology, similar to an export-based strategy, without benefiting from the relative labour-intensity and exploitation of economies of scale that an export-orientated strategy permits. Thus an export-oriented strategy would be likely to lead to a more appropriate technology than this type of import substitution.

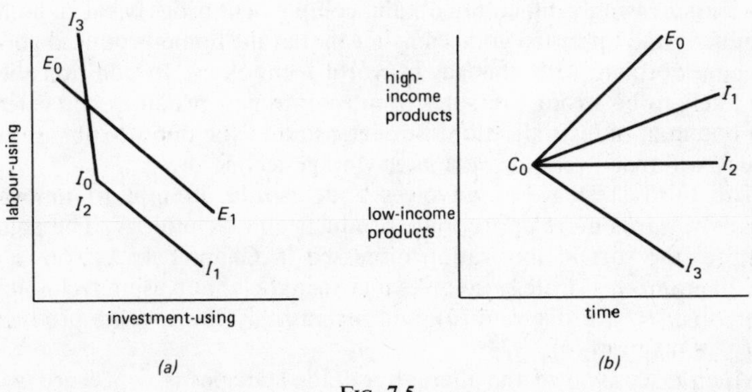


FIG. 7.5

Figure 7.5 illustrates the alternatives. Suppose $I_0$ and $C_0$ represent the initial policy, one of import-substitution. Diagram (*a*) shows the factor use associated with it, and diagram (*b*) the nature of the products consumed. An export-oriented strategy would involve a movement to $E_0$ leading to immediate employment gains, but subsequently movement along $E_0E_1$ would be necessary to keep up with world competition, involving a more investment-using and less labour-using technology. The export strategy would involve increasingly high-income products as shown on diagram (*b*) along $C_0E_0$. How this compares with an import-substitution policy depends on which strategy is adopted. The diagram illustrates three possibilities: $I_1$ which attempts to keep up with world innovation. This requires continued flow of technology from the advanced countries, and association with multinational companies. The strategy involves increasingly inappropriate technology, as shown by the line $I_0I_1$: the technology is *always* more investment-using than the alternative export strategy. The consumption patterns are assumed to move in the same direction as the export strategy, following advanced-country innovation, but with a bit of a time-lag to allow for the delay in transferring technology of the latest products.

The second import-substitution strategy, $I_2$, is one initially based on the same import reproduction as $I_1$, but subsequently no attempt is made to keep up with world trends. On diagram (*a*) technology use remains at the initial position, and similarly consumption patterns remain the same over time, as shown on diagram (*b*). Thus this strategy, while initially involving a more inappropriate technology than the export-promotion strategy, is soon overtaken by it, as the need to keep up with world technological advance dictates. Similarly, the consumption pattern remains at a certain level of inappropriateness but does not get worse. However, this strategy is very hard to maintain: changes in world technology will make the initial technology increasingly inefficient, parts will be increasingly difficult to obtain, complementarities between home-produced and imported goods may mean that the home-produced goods become obsolete with changes in world technology. In addition there are likely to be strong pressures to move into new products and extend the potential industrialisation. But extension of the import-substitution policy will mean replicating a later-vintage technology.

The third strategy, $I_3$, involves a deliberate attempt to innovate locally towards more appropriate products and technology. The policy requires the sort of innovation discussed in Chapter 4. As shown in the diagram, this strategy involves increasingly labour-using technology over time, $I_0I_3$ in diagram (*a*), and increasingly low-income products, $C_0I_3$, in diagram (*b*).

Much discussion of the alternative trade stategies is concerned with the immediate consequences, or comparisons of position $E_0$ with position $I_0$ on Figure 7.5 (*a*). In so far as a dynamic element is introduced, the comparison (implicitly) is between $I_0I_1$ and $E_0E_1$, where, as shown, the export strategy is continually more advantageous. Again, in so far as comparisons of consumption are introduced, it is assumed that both strategies would involve similar consumption patterns. But, as shown above, the conclusions are quite different if the comparison made is with strategies $I_2$ or $I_3$. $I_2$ is difficult to maintain, as argued, and is therefore likely to be transformed into one of the other strategies before long. Therefore the main alternatives are $E_0E_1$, $I_0I_1$ and $I_0I_3$.

Apart from the many difficulties in getting successful local innovation under way, a major objection to the appropriate import-substitution strategy is the high cost of autarky, given the economies of scale in innovation and production. But the dichotomy between free trade and autarky is a false one. The individual country is, in terms of economics, an arbitrary unit: there is no rationale in supposing that it is correct that there should be free movement of goods within it, but not outside. It is equally arbitrary to assume, as free traders do, that the world is the correct size for a trading unit. Some of the many intermediate arrangements may give the best combination of benefit from trade via economies of scale and specialisation, combined with sufficient protection from the outside world to secure the dynamic/development effects of protection. While it is true that neither theory[35] nor practice – as shown by the many trading arrangements made by groups of countries – have stuck to the rigid dichotomy between free trade and protection, there has been a tendency to regard such arrangements as second-best. Viner's classic work on Customs Unions, for example, regards such unions as desirable in so far as they are a step towards the freeing of international trade; the net balance between trade creation and trade diversion which he regarded as the test of the virtue of such unions (in which, with various qualifications about measurement and effects, he has been followed by most other theorists) was essentially a test of the extent to which a union contributed towards freeing trade, as against freezing it. Since most of the discussion on Customs Unions has been concerned with *allocative* effects, not *dynamic*, it does not, in general, provide for the possibility that a Customs Union (or other less than free-trade trading arrangement among partners) may have dynamic benefits for the countries concerned, which might justify the arrangements even if free trade were an option.

### THE CASE FOR SOUTH/SOUTH TRADE

Trade between third world countries would permit the development of appropriate technology on a third world basis. It presents the possibility of gaining potential scale economies without being forced to keep in line with advanced-country technology, in either production or consumption. In theory, trade between third world countries could consist in the production and sale of appropriate products using appropriate technology. If the third world as a whole protected its markets against North products, products sold within the third world need not use advanced-country technology, but could be adapted towards third world needs, enabling strategy $I_3$ above, an appropriate import-substitution policy, to be realised.

There is a case for reorientation of trade in a South/South direction simply in order to offset the many historic biases that have been established in the opposite direction.[36] But there is also a case for positive bias. The process of cumulative causation benefits the North in trade between North and South. Inappropriate (and expensive) technology is an inevitable feature of continued emphasis on North/South trade in goods and technology. On the other hand, autarky is not possible for many underdeveloped countries because of their small size, and the economies of scale.

Terms of trade considerations also tend to favour South/South links: unequal exchange, as described by Emmanuel (1972), arises because of the inequality in wage-rates between advanced and poor countries, so that one hour's labour in the advanced countries buys many more hours (say 5 to 15) of labour of the poor countries. The description of this as *unequal exchange* has its origins in Marxist terminology[37] – neo-classical terminology would reject the description so long as the wage-rate is equal to the marginal product of labour.[38] Nor does the fact that exchange is unequal in an Emmanuel sense have any relevance to whether there are or not gains from trade. The inequality strikes a metaphysical yet sympathetic chord: a movement towards South/South trade would remove this aspect of exploitation via trade. Prebisch's view of trends in the terms of trade is concerned with the changing terms of trade between primary products and manufactures following technical progress. He argues that the determinants of price are such that the producers of manufactured goods tend to capture the gains from technical progress by maintaining price and raising factor payments, while primary producers tend to pass on the gains from technical progress to the consumers of their products via increased supply and reduced price. Changes in the balance of supply and demand for primary products, and organisation of some primary product sellers, has perhaps reduced the validity of the Prebisch view for the relative pricing of primary products and manufactures.[39] But the view contains a kernel of truth in relation to the relative pricing of manufactures in trade between advanced countries and underdeveloped countries. This is partly a question of the relative wage/price relationship and partly of transfer pricing and technology payments. Workers in developed countries are generally speaking better organised, in a stronger bargaining position (because there is less surplus labour) and therefore are in a position to gain a larger share of a given product, and to capture a larger part of any increase in productivity, than workers in underdeveloped countries. The smaller share of wages in underdeveloped countries represents a poor deal for the country as well as the workers, because a considerable portion of the industry is owned by the developed countries, and consequently a corresponding share of the resulting profits is remitted there. To the extent that the stronger position of workers in developed countries allows them to capture the fruits of technical progress in higher real wages proportionately more than workers in underdeveloped countries can, the terms of trade between manufactures produced in developed and those produced in underdeveloped countries may move against the underdeveloped countries. Transactions within multinational companies, which, as Lall has emphasised, represent a considerable portion of world trade in manufactures, together with the explicit payments for technology and management services, enable multinationals to ensure that a large part of any gain in output resulting from technological progress is distributed to multinational companies in developed countries. This, rather than the question of primary products versus manufactured products, now seems to be a major source of deterioration in terms of trade between North and South, where terms of trade are broadly interpreted to include technology payments. A switch to South/South trade would reduce the scope for transfer pricing operations between North and South. However, it would only eliminate this aspect of deteriorating terms of trade if technological dependence were simultaneously reduced. A reorientation of trade is a necessary condition of reduced dependence, and therefore of eliminating the unequal and deteriorating terms of trade derived from technological dependence.

There are, however, some difficulties in relation to a move towards South/South trade: first, a switch to South trade from North trade may initially involve a switch to less efficient products and processes. This arises because of the technological dominance of the North and its ability to exploit economies of scale. Such cost is likely to be short-run because the switch should enable the South to develop efficient and more appropriate products and processes. In any case, for many countries there may be short-run gains to offset the short-run losses, in the form of market outlets for their products in other underdeveloped countries. This, of course, assumes (for which there is abundant evidence) spare capacity in industries in the South. But, given the similarity of pattern of development in many countries, spare capacity is often present in identical and not complementary industries. The more developed countries are likely to have spare capacity in industries (e.g. capital goods industries) for which there is a net import demand from other underdeveloped countries.

The different stages of development of underdeveloped countries presents the second major difficulty. The tendency for development to polarise, discussed above in terms of North/South development, is likely to occur between different underdeveloped countries if the more backward areas have no way of protecting themselves against the faster-developing areas. The gains from more South/South trade may be unfairly distributed initially, and the dynamic gains may be cumulatively unfairly distributed.[40] Therefore, any system for promoting South/South trade needs to embody some way in which the losing areas can protect themselves against the gaining areas, and some way of securing a fair distribution of gains. Although fair distribution of gains is obviously an important aim, the essential feature of any arrangement must be that no area loses *absolutely*: this is consistent with an unfair distribution of gains.

Thirdly, historical trading conditions have been responsible for the growth of North/South transport, communications, marketing arrangements, etc., that do not exist on a South/South basis. It is well-known that until recently in order to travel from West to East Africa one had to pass through London or Paris. Communications are even worse between different continents. Such trade as there is on a South/South basis is thus mostly confined to one region, and is not interregional, as shown in Table 7.2 above. The absence of transport and communications imposes a major obstacle to trading ties. This is a vicious circle because until trading ties are formed, improved transport and communications will not appear justified.

Fourthly, the economies of countries in the South have been (and are) dominated by the North/South connection to such an extent that it is extremely difficult for them to break out. For example, marketing and production franchises have been given to firms from the advanced countries. Technology purchase agreements include clauses restricting exports to third countries and tying imports to the advanced country.[41] Export credits are available for imports from the North. Tied aid requires purchases from the North. Links with advanced-country firms have created powerful vested interests in the South, among politicians, civil servants, businessmen, and the army, who resist any change in direction. Apart from the specific vested interests, the pattern of development arising from trade, capital and technology flows with the advanced countries has been responsible for an oligopolistic market structure and an inegalitarian income distribution which together reinforce North/South ties. The existing market structure and income distribution generate demands for the latest products and technology from the North, which makes any appropriate local technology difficult to establish. Thus, even in those countries which have technological resources and have developed some viable technologies, foreign technology continues to be imported.

All this means that even in one country, let alone the whole of the third world, a policy of orientation away from North/South to South/South trade, with associated changes in products and income distribution, would be difficult to achieve. But the problem is that the policy needs to be adhered to quite rigorously for the major gains to be realised. For example, suppose one country did reorient exports and imports towards other third world countries. If the other countries continued to use the latest technology and produce the latest products, the change in policy would change neither the nature of products nor the technology imported. Similarly, if the other countries continued to import freely from the North, then the country which had readjusted trade would have to compete with goods from the North, against a structure of consumer market that generated demand for the latest North-type product. Hence, to be successful, the country would have to import North technology, and would not be able to use appropriate technology and sell appropriate products.

The arguments are primarily concerned with trade in manufactures. In general, neither the cumulative causation argument nor the appropriate technology argument applies to primary products. For manufactured goods there are also exceptions, where continued trade between South and North would be advantageous. These include, for example, the processing of primary products.

### INSTITUTIONAL ARRANGEMENTS

The discussion of trade concluded that there were considerable advantages for underdeveloped countries in increasing trade links between third world countries, while reducing the orientation towards the advanced countries. A similar conclusion was reached in discussing ways of reducing technological dependence (Chapter 5), and the role of the capital goods industries (Chapter 6). However, despite the advantages there are powerful divisive influences. The historical near-exclusive links with the imperial country have been replaced by neo-colonial ties. Arrangements such as the Yaoundé agreements and the Lomé convention with the European Economic Community continue to stratify the third world on the old North/South basis. The difficulties experienced even by small regional groupings[42] suggest that arrangements involving the third world as a whole are likely to founder. On the other hand it is possible to devise aggregate schemes which would encourage third world trading links, without requiring a great deal of on-going cooperation. The rest of this chapter considers some of these schemes.

Possible arrangements include:

A Trade negotiations involving reduced tariffs on inter-South trade. These include arrangements confined to tariffs, and those involving overall coordination of economic policy.

B Monetary reforms which effectively favour South/South trade as against North/South. These include formal monetary unions, exchange rate policy, and the issue of special South international money to finance South/South trade.

In theory one can devise arrangements, classified under any of these headings, which would have equivalent effects. (E.g., a change in tariffs can be devised which is theoretically equivalent to a change in exchange rates.) In practice there tend to be important differences between the different policies in terms of negotiating problems presented, administrative requirements, the likelihood of countries breaking away from the arrangements, and the ease with which countries may safeguard their special interests. From the point of view of the earlier discussion there are certain features which we should look for in institutional arrangements. These are (i) that the arrangements should cover as many South countries as possible, and not be confined to countries in a particular region. From this it follows that policies requiring too much political cooperation of a long-term nature are not suitable. This rules out, for example, fully-fledged customs unions, at least initially. (ii) The arrangements should permit the least developed countries to safeguard their industries and their terms of trade, if necessary, against the more developed. Thus if the arrangements involve multilateral abandonment of trade restrictions (tariffs and non-tariffs), the less developed should be permitted either to go slow on these policies, or to have the effects in part offset by special financial arrangements (e.g. subsidies). (iii) The arrangements should allow countries which consider that the benefits outweigh the costs to continue to put their main efforts into North/South trade; this applies particularly to countries which are and will remain mainly sellers of primary commodities.

These conditions effectively rule out most trading arrangements (A above). As argued, there are two types of trading arrangements: those simply confined to tariff arrangements (and non-tariff barriers); and those involving a step towards full economic union. The former type of arrangement, which has been tried out to a small extent on a South/South basis,[43] makes no provision for the weaker areas to safeguard their developments, nor does it include provision for inter-country subsidies. Hence it might tend to weaken the less-developed areas, and for this reason is unlikely to be effectively introduced, or, if introduced, adhered to. On the other hand, more fully-fledged Customs Unions do normally include some provision for protection of the weaker areas; but they require a degree of political cohesion that makes them difficult to extend beyond a relatively small politically cohesive area.

Trading arrangements have been treated as *alternatives* to monetary arrangements. They are alternatives in the sense that one can achieve what the other could, if properly designed and executed. But they are also *complements* in the sense that one may support and encourage the other, and the use of one by no means rules out the other. The complementary nature of the two is recognised in many economic unions between countries, which normally require both tariff and monetary cooperation. Trading and monetary arrangements may be in conflict, one offsetting the effects of the other. For example, exchange rate changes may induce trade of a certain kind, but this can be offset by tariffs or other trade restrictions. Similarly, tariff concessions may be effectively negated by monetary arrangements. Consequently, one should not treat monetary and trade arrangements as independent, but consider them together.

Monetary reforms which would favour South/South trade include (i) credit unions between underdeveloped countries, so that payments between the countries are eased administratively, and credit extended between the countries; (ii) the issue of a form of international paper money to be used in (part) payment for goods and services from developing countries;[44] (iii) exchange rate policy such that the South as a whole acts as a block, with a joint devaluation against the North.

Formally each of these alternatives can be designed to achieve the same effects in terms of incentives for trade expansion among South countries. They need not be alternatives, but may be combined. We shall examine each, in somewhat more detail.

(i) *Credit union among South countries.* The first purpose of such arrangements is normally the easing of payments between the different countries, where lack of convertibility of currencies is inhibiting trade. This was the main feature of the European Payments Union. Lack of convertibility is not a major problem among South countries, as it was in post-war Europe;[45] in so far as it is a problem among South countries, such a union would ease it. The second purpose of payments unions is to extend credit among the members, enabling them to expand trade among themselves without risking the loss of (scarce) convertible currencies. Many variations are possible: 100 per cent credit, up to some limit, may be extended, or some proportion of debts may be met by credit. The net effect is to provide a considerable incentive for countries who are likely to be in deficit with other South countries to expand their purchases from them and meet the deficit with credit. But other countries, which are in surplus with South countries, but in deficit with North, may not wish to sell their goods in exchange for such credit. They require North currency to finance their North deficit. If the credit is only allowed in part-payment for the deficit with South countries, then sales to the South will earn these countries some convertible currencies, usable in the North. In addition, there may be arrangements whereby unused credits may be sold (to the central organisation or to deficit countries) in exchange for convertible currencies, at a market clearing rate.

(ii) *The proposed scheme for the issue of international money* – for use among South countries is very similar to a credit union. The new monetary unit is to be issued in fixed quantities, distributed among the countries in accordance with some formula (in proportion to the value of trade, for example) to be accepted in part-payment for goods and services from other South countries. The part-payments aspect means that countries do earn some convertible currency if they run up a surplus with the other South countries. Again some system of recycling the new unit from countries which have accumulated them to countries which have sold theirs, in exchange for convertible currency, is required to prevent the scheme seizing up, if there are some countries which are in permanent surplus with the other countries.

(iii) *Joint devaluation policy*. In this policy, all South countries would operate a common (devaluing) exchange rate against the North. This would have the effect of increasing the relative prices of North against South goods, thus presenting an incentive for the South countries to buy from each other, rather than the North. They would, at the same time, have an increased incentive to *sell* to the North. The scheme resembles the other schemes since the North-currency content of any given sales to other South countries would decrease relatively to that of sales to the North.

Despite the similarities, there are significant differences between the three schemes. In the first place, the problems of administration differ. A credit union involves continuous administration, whereas the new monetary unit may be issued once and for all. The joint devaluation scheme is similarly once and for all. But there are administrative problems – in particular effective mass devaluation against North currencies can only be achieved with the cooperation of the North, whose buying and selling of currencies may determine the exchange rates. There is a second major difference between the credit/monetary unit schemes and the devaluation scheme. The latter operates directly on prices paid and received by traders. The former operate on the incentives of those responsible for monetary and economic management. They would have no immediate effect on prices traders face, and would only affect trading patterns to the extent that governments pursued tariff, direct purchases, and bilateral trading policies favourable to inter-South trade, as encouraged by the scheme. In a sense the main incentive effect of the schemes would be on the economic managers; the managed would only be affected via subsequent changes in policy. This would have the advantage that governments could operate the incentive towards South trade selectively, picking out, for example, manufactures, where the main arguments for trade expansion arise, and leaving the export of primary commodities unaffected.

As argued above, the schemes need not be alternatives; the most desirable solution might be a combined scheme. A credit union could be accompanied by the issue of paper money; and both could achieve their effect by exchange rate (and/or tariff) changes.

It has been argued that factors that inhibit South/South trade are not monetary at all but tradition, lack of market contacts, lack of transport links, and poor-quality goods. In part, clearly all these factors are responsible for the small amount of South/South trade. Also, as argued earlier, technological backwardness means that the North captures all the technological gap trade. Monetary reforms may none the less be effective in producing an expansion of South/South trade, even if the explanation of low levels of trade is not to be found in monetary institutions, since financial incentives can compensate for other disadvantages; thus poor-quality goods may be a best buy if they are cheap enough. Expensive and cumbersome transport routes may also be compensated for. What cannot be compensated for is the total absence of a crucial link – e.g. where there is *no* method of getting the goods from A to B then financial incentives will not make a difference. But the way this has been posed shows that it is not a likelihood. If goods can be transported from A (South country) to C (North country) and from C to B (South country), then goods can go from A to B via C and, given sufficient financial incentive, they will do so. In any case factors such as transport cannot be taken as givens, except in the very short run. They are the outcome of trading ties. This is an externality of trading which means that there is a tendency for trade to expand in traditional channels. Routes between A and B will only be developed if there is seen to be potential trade between A and B. If the absence of routes is taken as an argument against developing trading ties, then the *status quo* trading situation will always be seen to be justified, even though an assessment of the relative advantages of A–B and A–C trade, allowing for the transport and other links that will develop with the trade, would argue for the creation of A–B routes *and* trade.

It was argued above that a necessary feature of any scheme was that it protected the weaker areas. Some trading schemes do not do this. But the monetary schemes do, in so far as they allow the weaker areas to protect themselves, by tariffs and in other ways, and by action on the exchange rate. The joint devaluation scheme, though in theory it permits countries to vary their exchange rates relatively, would in practice be difficult to combine with the kind of freedom over exchange-rate changes that countries prefer and need. For this reason, the other schemes seem more likely to be adopted. This is particularly likely to be true for countries with long ties with particular North countries, who might refuse to devalue against these currencies, but for whom extension and receipt of credit with other South countries would be more acceptable.

In conclusion it must be emphasised that whichever approach is adopted, initially, it should not be regarded as exclusive. It may fruitfully be combined with other aggregative approaches to greater South/South trade, monetary, trading or institutional (e.g. new transport links). Policies of industrial coordination may complement and support or conflict with trading arrangements. In the Latin American Free Trade Area multinational companies have supported monetary and trading arrangements so as to coordinate their industrial policy. Elsewhere, as in East Africa, apparently significant agreements have foundered on lack of industrial coordination. Aggregate approaches are perfectly consistent with further moves towards regional economic groupings. In fact the more successful the regional economic groupings the easier such schemes would be to introduce, since the number of independent parties concerned would be that much smaller.

NOTES

1. The term Heckscher-Ohlin is used here to cover the corpus of neo-classical trade theory derived from Heckscher and Ohlin and from developments such as those of Stolper and Samuelson. This does not mean that the theory, as described, would necessarily have been recognised or accepted by its originators. It has been argued that it departs considerably from the spirit of Heckscher's writings, in particular in the technology assumptions. Johnson (1968) describes the assumptions as being aspects of 'The more ambitious contemporary Heckscher-Ohlin model . . .'.
2. Since it is normally assumed that there is a wide technical choice for the production of any good, it may not make sense to speak of labour-intensive (or capital-intensive) *goods*, for each good *may* be produced in a capital-intensive or a labour-intensive way. If technological possibilities are such that any good can be produced with any ratio of capital to labour, and equally efficiently from a technical point of view, then differences in factor availability would not explain international trade, and other reasons for differences in comparative efficiency would have to be found to explain the advantages of trade. However, the empirical findings of similar ordering of goods in terms of factor requirements in different countries (see the discussion in Lary (1968), Chapter 3), which has been described as the absence of factor-intensity reversals, justifies the description of goods as capital- or labour-intensive.
3. Economic conditions clearly do influence the location of production of many primary commodities: it is rare that the only determinant of location is nature.
4. By Hirsch (1973(b)).
5. According to developments of the H-O theory by Stolper and Samuelson.
6. 'The introduction of factor categories, such as "technical labor", i.e. capital-intensive labor, is a gimmick which, when its superficial advantages are more closely examined, turns out to rob the factor proportions theory of all meaningfulness.' Burenstam Linder (1961), p. 86.
7. Most of the evidence relates to GNP per capita (see Kuznets (1972)). In so far as real wages have been increasing faster, in LDCs, than total incomes the wage gap between organised workers may have been closing. But we are far from factor price equalisation.
8. These are rigorously described in J. Bhagwati (1969), and H. G. Johnson (1958).
9. Johnson (1968), pp. 9–10.
10. Formally, the theory of comparative advantage may be adapted to allow changes over time by redefining it as 'dynamic' comparative advantage. This has been done, e.g. by Johnson, leaving the essence of the theory, and its conclusions, unchanged. Thus, as with production theory, changeless 'change' incorporates change theoretically while leaving the static framework and conclusion essentially unchanged.
11. There is abundant empirical evidence for the existence of increasing returns in industrial technology. Some of it is summarised in Chapter 3.
12. There are implicitly two theories here. One relates the rate of growth to the absolute size of the initial base: this could be explained by economies of scale in growth-producing activities, e.g. R and D, or by the fact that the size of growth-related activities is positively related to the size of output (e.g. savings proportion and R and D as a proportion of total income). The other theory relates the rate of growth to the past rate of growth. This could result from a particular type of

technical progress, positively related to the rate of growth of output and the capital stock.

13. See Griffin (1974).
14. Ibid.
15. Freeman (1968) shows, for example, that exports in chemical process plant were usually based on a large home market: Hirsch (1967) provides similar evidence for electronic firms in the U.S.
16. This is the process described by Posner (1961 and 1970) and Hufbauer (1966).
17. An explicit assumption of the Posner model was *equal factor endowment* in the trading countries. But Hufbauer also allowed for low-wage trade flowing from poor to rich countries, as in the product cycle theories.
18. Of Vernon (1966) and Hirsch (1967).
19. A process described by Helleiner (1973) and Sharpston (1975).
20. The concept of product efficiency is discussed in Chapter 1.
21. This *ceteris paribus* covers a host of conditions. Innovation in the South might not consist of more appropriate products if the scientific, technical and economic system in the South supported North-style innovation and markets. See Cooper (1974), *passim.*
22. For evidence of income distribution in some underdeveloped countries, see Adelman and Morris (1973) and the IBRD/IDS report by Chenery and others (1974).
23. See Chapter 3 for an expansion of the argument.
24. A concept first used in relation to United States/W. European trade, by Stolper (1950).
25. See, for example, Lary (1968), Hufbauer (1970) and Hirsch (1973(b)).
26. D. H. Robertson summarised the position in a famous parable (describing U.S./U.K. trade in the early 1950s):

    The simple fellow who, to the advantage of both, has been earning a living by cooking the dinner for a busy and prosperous scientist, wakes up one day to find that his master has invested in a completely automatic cooker, and that if he wants to remain a member of the household he must turn shoe-black. He acquires a kit and learns the techniques, only to find that his master has invented a dust-repelling shoe, but would nevertheless be graciously willing for him to remain on and empty the trash-bins. Would he not do better to remove himself from the orbit of the great man and cultivate his own back garden? And if he can find some other simple fellows in the same boat with whom to gang up and practice the division of labour on a less bewildering basis, so much the better for him.

27. Managerial requirements are related to *change* – see Penrose's analysis.
28. It is common for those advocating free trade to qualify their advocacy in minor ways: e.g. Little, Scitovsky and Scott favour a 5 per cent subsidy on manufacturing industry.
29. A term developed by K. Marsden of the I.L.O.
30. Such consequences have been widely shown empirically – see e.g. Khan and Islam (1967) on Pakistan, and Little, Scitovsky and Scott, *passim.*
31. As argued by Burenstam Linder.
32. See Subrahmanian (1972).
33. Some of these are described by Vaitsos (1974), and Lall (1973).
34. Furtado (1964).
35. See the voluminous literature on Customs Unions, e.g. Lipsey (1970), Viner (1950) and Robson (1971).
36. Some evidence of this for W. Africa is included in Amin (1973) and Hopkins (1973). According to a report on South America, 'the national transport systems served in the past to channel the production of raw materials to the ports for export to the industrialised countries, and only recently has there been

super-imposed a system designed to link the main population centres with one another within each country.' *Economic Bulletin for Latin America*, Vol. XVIII, Nos 1 and 2 (1973).

37. Though some Marxist theorists have denied its existence (see Kidron (1974)) on the grounds that subsistence requirements are higher (in proportion or more than in proportion to the higher wage-rates) in developed countries.
38. Griffin (1974), oddly neo-classical in this respect, defines unequal exchange as occurring where the ratio of price to marginal cost differs between those trading.
39. The view has been much criticised both theoretically and empirically – see, e.g., Meier (1968).
40. Analysis of CARIFTA trade (1967–70) shows that most of the expansion in trade was attributable to trade between the four largest countries. See *Economic Bulletin for Latin America*, Vol. XVIII, Nos 1 and 2 (1973) pp. 139–49.
41. There is abundant evidence of these practices: see Vaitsos (1974), Lall (1973) and UNCTAD, TD/106, TD/B/AC 11/10.
42. See, e.g., Dell (1963) and Maritano (1973). Customs Unions even among such similar countries as those of the East African Common Market have met with formidable obstacles, both economic and political.
43. In 1971 a GATT protocol established mutual trade concession between sixteen developing countries. See GATT Press Release GATT/1097, December 1971.
44. The proposal was described in Stewart and Stewart (1972).
45. Though it does undoubtedly impose some obstacles to trade. Thus trade among the Maghreb States has to be financed via francs because of the inconvertibility of the States' currencies. See Robana (1973).

# 8 The Choice of Technique: Empirical Studies

The earlier parts of this book suggest the areas which require empirical illumination. Chapter 1 presented a general model of technological choice: each technique is associated with a vector of characteristics; the set of techniques available depends on the historical development of technology, and specifically the economic/social conditions of the economy for which the techniques were originally developed; each decision maker has certain objectives, a certain amount of knowledge about the technological possibilities, faces certain restraints, and controls a certain amount of resources. The actual choice made then depends on the interaction between decision makers and technological possibilities, given the various objectives and restraints. Ideally, empirical studies should try and illuminate this complex process – revealing for example how the historical development of the industry affects the technology available; *whether* and *how* one set of decision makers, e.g. multinational firms, make a different choice from another, e.g. local publicly owned firms; and *why* – e.g. because objectives differ, or because access to resources differ, or because markets differ. Chapter 4 presented a more specific empirical question: that is, the question of the existence and relative efficiency of an appropriate technology, and the sort of socio-economic changes that would be needed to get such a technology – if it exists – into use. This question is in a way a sub-question of the general model of technological choice in Chapter 1.

Because most of the discussion of choice of technique has centred around the neo-classical model, as described in Chapter 1, most of the empirical studies have been similarly *primarily* concerned to determine the neo-classical versus technological determinist debate – viz. whether the neo-classical assumption of a near-infinite array of different techniques to produce a given product is correct, or whether because of the dominance of recent technological developments there is only one efficient technique available for any given product. Secondly, studies have been concerned with the surplus maximisation question – viz. given the assumption of a neo-classical choice of technique, which technique maximises the surplus available for reinvestment.

While empirical studies concerned with such questions also shed some light on the more general model of technological choice presented here, they do so only to a limited extent, because the terms of reference required to answer what, for shorthand, we shall term the neo-classical question automatically exclude many of the relationships. The most important type of exclusion arises because to answer the neo-classical question techniques producing the *same* product have to be compared (i.e. all techniques compared must produce the same quantity of a homogeneous product). But differences in product characteristics – quality, sophistication, materials etc. – provide one of the most interesting dimensions of technological choice, as discussed in Chapter 1. Moreover, one of the most significant changes in technology historically has been the increasing scale of production for which techniques were designed. Consequently another critical dimension of technological choice is the scale factor. Further, the neo-classical model assumes uniform access to technical knowledge among decision makers in terms of quantity and price, uniform access to resources, and uniform motivation, thus automatically ruling out study of how the choice made differs.

Studies directed at the neo-classical question often also have limited relevance to the appropriate technology question, and for much the same sort of reason. As discussed in Chapter 3 the characteristics of more appropriate technology differ from those of advanced-country technology in a number of respects – in particular, the nature of the product differs, being simpler, making more use of local resources and being designed for low-income consumers, and the scale of production is likely to differ, being smaller and more suited for rural location. But the studies designed to answer the neo-classical question do not generally include these dimensions of choice. Hence, mainly they shed only indirect light on the question of appropriate technology; more empirical evidence is contained in the many examples that have been collected by the ITDG,[1] though these for the most part describe simple technologies, and do not provide a full comparison with alternative techniques.

Not all micro-studies have, of course, stuck strictly to the neo-classical question, and some therefore do illuminate the general relationships. The rest of this chapter reviews some of the results of empirical studies. The next two chapters report on case studies conducted in Kenya, which, it must be admitted, were also initially designed to answer the neo-classical question.

Broadly, two approaches have been adopted. First, the macro-approach of fitting production functions – of a constant elasticity of substitution (C.E.S.) type – to cross-section data within an industry, and sometimes between industries. This approach was pioneered by Arrow, Chenery, Minhas, and Solow (1961). The approach suffers from the conceptual defects of C.E.S. production functions. In general, vintages are ignored and machinery of widely differing age and origin is treated as if it were all part of a single production function. Capital is treated as a homogeneous input, and its marginal product is assumed to be shown by the profit rate (while similarly the marginal product of other factors, like labour, are also assumed to be shown by their rewards). Given these (and other) assumptions the production function which best fits data is calculated, and the elasticity of substitution between factors derived. The elasticities so derived are as much a derivative of the (artificial) assumptions made as of the data, and do not shed much light on current opportunities of developing countries. The approach is subject to systematic criticism by O'Herlihy (1972), and also by Harcourt (1972). Morawetz (1974 and 1976) shows the inconsistency between the industry rankings generated by different studies using the C.E.S. approach.

The second approach is that of the micro-studies. These look at the required inputs to produce a given output – normally a single product or very close substitutes. This chapter concentrates on these micro-studies. The first part considers some of the problems that arise in such studies; the second discusses some conclusions emerging from the micro-studies.[2]

## METHODOLOGY AND PROBLEMS

The recent severe and penetrating attacks on the concept of capital[3] might be thought to have discredited any studies designed to assess the capital-intensity of techniques right from the start. It has been shown that there is no objective entity *capital* which may be objectively measured in the way that one might count pebbles on a beach. The capital stock consists of heterogeneous items produced at different times, with different lengths of life. Summing these involves weighting, and the particular sum arrived at – the 'quantity of capital' – depends on the weighting used, and in particular on the time-weighting, or the rate of interest. Thus neither the quantity of capital, nor the marginal product of capital are objectively identifiable, which was one of the chief reasons for criticising the C.E.S. approach.

However, the attack is on the concept of *capital*, not that of *investment* which is the concern of micro-studies of choice of technique. These studies do not try to aggregate the value of past investments, but rather look at the current investment costs of introducing different techniques.[4] Some weighting is involved here because investment decisions too involve non-homogeneous items, over time as well as across time. With assets of different lives, the interest rate assumed will influence the values reached. Thus the investment costs of the techniques are not objective quantities, but are dependent on the various systems of weighting adopted. This does not invalidate the studies: they are necessary inputs for making investment decisions, and for realistic discussion of the theory of technical choice. But it does mean that the investment-intensity (and the other associated ratios) of any technique is dependent on the value of the weights adopted, and is not objective and unalterable. Some of the studies show themselves more aware of their tentative and non-objective status than others.

There remain serious problems in estimating the investment costs of different techniques; these derive partly from the general valuation problem facing any attempt to make economic estimates: should market prices be used or should these be 'corrected' for market imperfections?[5] Some studies use market prices in an underdeveloped country,[6] others use prices derived from the market prices ruling in the supplying (developed) country,[7] while yet others attempt to use corrected prices.[8] Comparisons between techniques also may differ according to whether the length of life of assets is taken into account, when the lives of the different assets differ radically. Again the comparison may be altered by the inclusion or otherwise of repair and maintenance costs. This is illustrated in the case studies in the next chapters. A further difficulty concerns what should be included in the investment costs. In some economies the efficient use of a particular technique may require all sorts of infrastructure-type investment which in other societies is already provided automatically. Doyle provides an example in his comparison of two more or less identical plants in the U.S. and Indonesia. The Indonesian plant (which was smaller than the U.S. plant) needed over twice as much investment to provide services (e.g. transport) that were already in existence in the U.S. case.

Then there is the question of working capital. This is often ignored in comparisons because it is notoriously difficult to estimate. But the omission may produce misleading results, since it is likely that working capital requirements vary systematically with the labour-intensity of techniques. Working capital arises because of the time-lag between expenditure on recurring costs (materials, fuel and wages) and receipts from selling output. Recurring costs are likely to form a greater proportion of total costs the more labour-intensive the technique.[9] However, this tendency might be offset by lower wage-rates paid to labour in labour-intensive activities, or lower non-labour recurring costs (e.g. no fuel costs). Some modes of activity, e.g. self-employment or family labour, may not require that wages are 'advanced' and may not involve this type of working capital. Prasad (1963) has illustrated the potential significance of working capital as shown in the table below.[10] However, it must be emphasised that it is incorrect simply to add working capital to fixed investment requirements because working capital is ultimately recovered while fixed capital is not. The 'costs' of working capital then depend on the time valuation – with a zero time preference, working capital would be costless.[11]

TABLE 8.1

| *Industry and method* | (1) *Fixed capital % of total inc. working capital* | (2) *Fixed capital per employee* | (3) *Total capital per employee* |
|---|---|---|---|
| | | (Rupees) | |
| 1 Rice | | | |
| A Pestle & Mortar | 1·8 | 5 | 268 |
| B Ordinary dhenki | 5·0 | 16 | 316 |
| C Improved Assam dhenki | 1.4 | 19 | 1370 |
| 2 Sugar | | | |
| A Cottage gur industry | 25·0 | 126 | 503 |
| B Cottage Khandsari | 29·0 | 516 | 1781 |
| 3 Cotton spinning | | | |
| A Ordinary Charkha | 20·0 | 10 | 50 |
| B Ambar Charkha | 33·5 | 50 | 150 |
| 4 Cotton weaving | | | |
| *Hand* | | | |
| A Throwshuttle | 1·6 | 4 | 243 |
| B Flyshuttle | 8·1 | 32 | 390 |
| C Banaras semi-automatic | 11·7 | 133 | 1133 |
| D Madanpura semi-automatic | 10·0 | 167 | 1667 |
| *Power loom* | | | |
| E Small-scale, cottage industry | 64·0 | 3448 | 5388 |
| F Non-automatic power loom | 39·0 | 7976 | 20,450 |
| G Automatic power loom | 57·0 | 159,600 | 280,000 |
| 5 Soap | | | |
| A Cottage | 18·0 | 163 | 910 |
| B Handmade soap industry | 45·0 | 1865 | 4125 |
| C Small-scale industry | 40·0 | 1480 | 3700 |

SOURCE K. Prasad (1963), Table 80.

*Output*

Estimating output related to different techniques raises two types of question: first, the nature and quality of the product; and secondly that of rate of output and capacity utilisation.

*Type of output:* the earlier discussion concluded that an important dimension of choice of techniques is choice of product. Many different products fulfil the same need. Such products may differ in most respects – for example, sausages and milk both fulfil a need for nourishment – or in a few respects, where the differences are usually called 'quality differences': finely spun and coarsely spun cotton both fulfil the need for thread. The extent to which different products or products of different quality do fulfil the same need depends on how the need is specified. The more generally needs are specified the greater the number of products of varying quality that may fulfil them. The more narrowly specified the fewer products: if very narrowly specified only one product may fulfil the need. For example, only Renault 16 car doors will do if one needs to replace a door on a Renault 16 car. Techniques of production very often differ in the quality of products produced, and they may also differ in the type of product produced. This raises two difficult questions. First, any study of techniques has to start by deciding which techniques are to be included. To do this it is necessary to decide how the needs are to be specified and hence the variety of products that would fulfil the needs. Secondly, since the different techniques are associated with different products, the study has to value these differences in coming to results.

A needs-based approach to choice of technique threatens to encompass a wide variety of different products if needs are broadly defined – thus raising the major difficulty of how to value such differences. On the other hand, excluding product variations, by making special efforts to find alternative techniques for producing a homogeneous product, as many studies have done to avoid valuation difficulties,[12] restricts the comparison so that little choice of technique is likely to remain, while it begs the main question – that of the variety of possible ways of meeting given needs. The two case studies that follow indicate how relatively minor product differences may be critical in determining choice of technique.

Economists have suggested two ways of valuing different products: a simple way round many of these problems is to take the consumers' sovereignty way out, valuing products according to market prices.[13] But the conditions required to make this valid – desired income distribution, perfect competition, no externalities, no advertising – are so far from being met that consumers' sovereignty is a snare not a solution. The second method suggested[14] for open economies is to take export price as a guide to the value of output, since this represents foreign exchange that might be acquired. This is legitimate where the items in question actually are exported. Where they are not, but satisfy previously unsatisfied domestic demands, act as import substitutes, or substitutes for other domestic output, the foreign exchange price is *not* a guide. To take the foreign exchange price to compare the value of e.g. mud bricks with concrete walls implies accepting the 'world' valuation of mud bricks and walls – a valuation which reflects world income distribution and taste patterns. Where home production acts as a direct and exact substitute for imports then import prices may be a correct guide to foreign exchange saved. But in choice of technique studies the central question is the different ways in which needs may be fulfilled. Exact import reproduction is only one way, and hence import prices often provide no guide to relative valuation of different qualities of goods.

One way of approaching the problem is to try to specify the need that is to be met as precisely as possible. Any output that meets this need may then be classified as of the same value, for the purpose of the exercise. If it overfulfils the needs – producing e.g. stronger bricks than specified – this should not add to value, at least in the initial assessment. If it can be shown that this overfulfilment – or meeting of non-specified needs – is of value, then this must be weighed in the final conclusion. In a study of can sealing a major reason why the automatic method was preferred was that less supervision was required to get the quality of cans necessary for export markets.[15] In so far as the cans are in fact going to be exported, then this extra quality presumably is necessary. But it may not be necessary for domestically consumed cans, or for exports within East Africa. There is an interaction between the results of the studies and needs: most needs are not autonomously defined but are related to the costs of meeting them. For example, if it is shown that two-storey building is going enormously to reduce employment and increase costs compared with one-storey, this might lead to a rethinking of the building programme in terms of one-storey accommodation. Hence *prior* definition of needs, to rule out single-storey accommodation and therefore study of hand-block makers, would misleadingly limit the scope of the study.

*Rate of output:* theoretical models of technical choice have been divided into three types[16] – putty-putty, putty-clay and clay-clay. The putty-putty model assumes that the investment/labour and investment/output ratio may vary *after* the investment takes place as well as before. There are thus no unique ratios – of $I/L$ or $I/O$ – associated with each technique. Such assumptions make nonsense of choice of technique studies, which aim to ascertain the ratios associated with each technique. The putty-clay model assumes there exists a range of techniques before the investment is undertaken, but once installed each machine is associated with a given rate of employment and output. The clay-clay model assumes that there is no choice of technique either before or after the investment is made. Thus choice of technique studies designed to answer the neo-classical question can be viewed as throwing light on whether the putty-clay or clay-clay models are nearer to reality. But survey data shows that output and employment do vary for each technique, as well as between techniques; put in another way there is some putty-putty element.

There are two main sources of output variation with any given technique: the rate of output associated with any technique may vary with the intensity of its use when in use, and the amount of time when actually in use, or its capacity utilisation. The studies of cement blocks and maize grinding which follow show the large variation in output per machine-hour associated with the same machine. Such variations make it difficult to talk of 'the' ratios associated with any given technique. Sources of variation include managerial efficiency, worker skill, demand for output, and the numbers employed. The range differs with machine design. For cement blocks the pace chosen depended, among other things, on the number of workers available to prepare the mixture, move the blocks, etc. Another factor determining output per shift is the amount of downtime necessary because of repairs. This is likely to vary with the age of the machine.[17] To the extent that older machines are also more labour-intensive they will show a greater amount of downtime. There is also the Hirschman machine-paced factor. This suggests that variations in the rate of activity are likely to be greater with less mechanised, and therefore less machine-paced, techniques.[18] Such variations in output not only make it impossible to discuss *the* $I/L$ or $I/O$ ratio associated with each technique; they also mean that studies based on a single case, or engineers' views about what performance ought to be, may be completely misleading. The point is that each technique is not correctly represented by a single point in an isoquant diagram, but consists of a range as depicted in Figure 8.1.

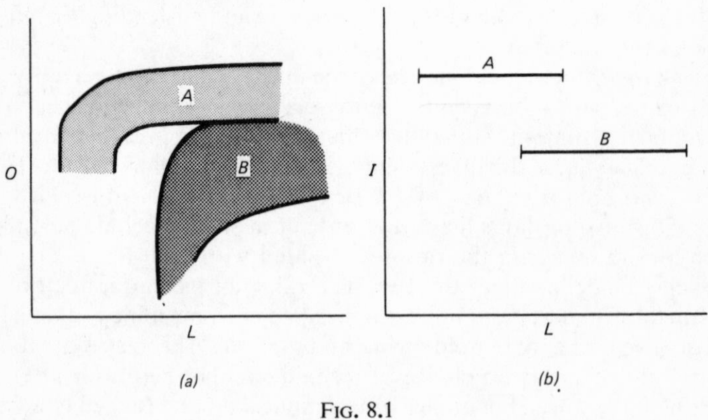


FIG. 8.1

Diagram (*a*) shows the range of productivity associated with two techniques, *A* (which is more investment-intensive) and *B*. For each technique output per machine-hour increases within a limited range, as employment increases, but more variation is possible for the labour-intensive technique. Diagram (*b*) shows how for a given rate of output, employment may vary for each technique. The relative efficiency of the two thus depends on how each is operated.

*Scale and capacity utilisation*: machines are indivisible. The ratios of both $I/L$ and $I/O$ depend on the degree of capacity utilisation. Often the more mechanised techniques are designed for larger-scale operations than older labour-intensive techniques. The comparison between techniques can be crucially affected by the size of the market. Pack (1974) found that: 'though there was almost universal agreement about the economic irrationality of automatic filling machines at low volumes, there was similar unanimity about its desirability at high volumes'. Similarly in the manufacture of cement blocks the hand-operated machines were the most sensible choice at low levels of output.[19] The indivisibility argument should mean that larger units are more likely to be operated at lower rates of capacity utilisation than smaller units, and hence investment-intensive techniques (which tend to be larger) at lower rates than labour-intensive. However, other factors may operate in the opposite direction. A defining characteristic of investment-intensive techniques is that their ratio of unavoidable to avoidable costs is higher than that of labour-intensive techniques. If two techniques have the same unit costs at full capacity, the rise in unit costs as capacity utilisation declines will be greater for the investment-intensive technique than for the labour-intensive technique.[20] Therefore one would expect more incentive for managers to operate investment-intensive techniques at fuller capacity, and thus some differences in rates of capacity utilisation in this direction.

Other factors may also influence capacity utilisation. Differences in wage rates, and also wage-rate differentials between night and day operations[21] alter the cost savings discussed above; also differences in the ratio of avoidable to unavoidable labour costs (which is related to the ratio of skilled and managerial to unskilled). Prasad found that small-scale techniques operated to low capacity because they could not afford to finance the working capital essential for fuller capacity utilisation.[22]

*Choice of techniques to study*

Any machine is normally only a small part of the picture. There are vertical and horizontal links with other investments in the productive chain. Horizontally, complementary investments are required such as buildings. Vertically, there are all the preparatory activities – preparation of materials, transport to the site; and all the subsequent activities of processing, packing and transporting that may not be covered by what has been described[23] as the 'core' technique. To the extent that there are technical and scale links between the different investments, then study of one element in the chain may give misleading results. For example, Pack suggests that the introduction of automatic receiving equipment in one firm in Kenya saved both investment and labour, and possibly reduced the investment/labour ratio since a smaller number of trucks were needed to collect the fruit, which would now be done in minutes rather than hours. The World Bank got similar results in looking at mechanical loaders in a road project in India.

The peripheral activities discussed so far are all, in a sense, on site. But each complex (including core and peripheral activities) is one stage in a production process. It takes inputs from other processes and its output forms the input of further processes. Just as 'core' choice of technique may be a small part of the total on-site choice, so site activities may be a small proportion of the total chain. In choosing techniques, the choice is not only of the technique at a particular link in the chain, but also of the chain itself. For example, we may compare two chains consisting of maize versus wheat.[24] Broadly the chains look like this:

| *Maize or wheat – chain* | *Rough comment* (on basis of Kenya experience) |
|---|---|
| preparing ground<br>planting<br>weeding<br>harvesting<br>dehusking | maize tends to be smaller-scale and less investment-intensive. |
| milling | wheat investment-intensive: greater choice of technique exists in maize. |
| cooking | wheat cooked in part commercially with some choice of technique, maize domestically; wheat higher nutritional value, but nutritional value depends on processing technology. |

The actual process of milling forms a very small part of the total chain. Looking at the chain as a whole might alter the conclusions arising from looking at one link.

*The surplus:* much of the literature on choice of technique pays considerable attention to surplus generation, and the ratio $S/I$ (or surplus generated per unit of investment) is regarded as of greater significance than output per unit of investment.[25] Surplus generated per technique is commonly identified as the difference between value-added and wages.[26] Consequently use of surplus generation as a criterion tends to favour less labour-intensive techniques with smaller wage bills. But the significance of the surplus, when identified in this way, depends on the mode of production prevalent. The surplus is relevant to the growth of output and employment in so far as it is reinvested in the economy. In a planned economy it can be assumed that all profits are saved, and available for reinvestment. But in a fully planned economy appropriate savings levels may be raised in other ways (for example, through pricing or tax policy). In private enterprise economies it cannot be assumed that the whole surplus is reinvested. In the private domestic sector, profits may be consumed so the relevant surplus is:

$$S = sp\,(O - w \,.\, L) + t \,.\, w \,.\, L$$

where $sp$ is the propensity to save out of profits;
$O$ value added;
$w$ the wage-rate;
$L$ number employed;
$t$ the propensity to be taxed out of wages.

Both the propensity to save out of profits and the wage-rate vary with the nature of the enterprise. While evidence about savings is scanty, there is considerable evidence[27] that there are substantial variations in wage-rates between large- and small-scale enterprises and urban and rural enterprises. Much lower wage-rates paid by small-scale activities using labour-intensive techniques may compensate for the greater labour input in calculation of surplus.

The existence of a foreign-owned sector also alters the relationship between surplus generation and investment. The foreign sector can determine its investment policy independently of its savings in the particular country, increasing or reducing its remittances so as to reconcile the two. Thus additional surplus generation in the foreign sector may simply mean greater foreign remittances, not greater investment. To the extent that the foreign sector uses local finance and/or competes with local sources of finance to finance the same projects, greater flow of local finance, through greater surplus generation, may reduce the net flow of foreign finance and may not affect the investment rate. Thus the nature of ownership of the assets is essential information in assessing the significance of rates of surplus.

*Income distribution effects:* these relate both to the distribution of earning opportunities and the nature and distribution of the resultant product. While theoretically distribution of income may be offset by government tax/subsidy policies, in practice the theoretical adjustments that might take place rarely do. Indeed one of the important reasons for desiring a more appropriate technology (see Chapter 4) lies in its distributional impact in contrast to inappropriate technology. Consequently, income distribution effects should form a central focus of the studies, though in practice they have often been ignored.

*Data sources*: there are three types of data. Engineering data derived from blue-prints supplied by engineers and costs from the manufacturers of the machines; data derived from detailed case studies of one or two techniques in use; survey data taking a number of examples of each technique.

As argued above, performance with techniques may vary between economies and within economies; such variations may be related to the labour-intensity of the techniques. Consequently, the use of engineering data alone can be most misleading.

Studies of the performance of techniques are in some ways time- and country-specific and the results cannot be generalised – time-specific because technological change renders the studies obsolete as fast as the machines; country-specific for two reasons: first, because technical requirements may vary between countries because of differences in geography, managerial skill, raw material resources. But even with identical technical characteristics, economic differences may lead to differences in investment and other costs. To calculate investment costs requires estimating and adding up non-homogeneous costs, including local labour of various types, locally produced and imported materials and machinery, which requires economic evaluation. While empirical case studies of the choice of technique generally try to stick to technical values as much as possible, all the studies, to varying degrees, make use of economic data as well. It must therefore be emphasised that apparently technical ratios comparing $I/L$, $I/O$ for different techniques are based in part on the conversion of non-monetary (technical) values into money values. This inevitably involves some assessment (implicit or explicit) of the relative value of different items, and consequently the ratios may be altered by a different relative valuation.

### SOME CONCLUSIONS FROM EMPIRICAL STUDIES

It is perhaps helpful to consider these in the light of the general discussion of technological choice contained in Chapter 1. This suggested that the range of available choice, in terms of labour and investment intensity, would depend on the historical development of the industry. Industries which were originally developed a long time ago are more likely to show a wide range of techniques than industries recently developed. In old industries the older techniques may sometimes still present efficient alternatives to recent techniques: whether they do or not depends on the extent of productivity increase associated with recent techniques. There is no *general presumption* that all old techniques have become technically inferior, though it is likely that some have, while those that have survived are likely to be fairly inefficient – i.e. with low labour and investment productivity – because they were developed long ago. Nor can there be a *general* presumption that all old techniques remain efficient. Hence it seems likely that the results will vary industry by industry.

The second general conclusion from the discussion in Chapter 1 was that the question of investment and labour intensity represents only one aspect of technical choice: other characteristics, including especially scale of production and product type and quality, will also vary, probably in a systematic way.

Most micro case studies of choice of technique have concluded that, given certain assumptions and within certain limits, there is a technically efficient[28] range of techniques of differing labour and investment intensity, in the industries studied, thus rejecting the technological determinist view and providing support for the neo-classical assumption. This is the conclusion, for example, of most of the Strathclyde studies,[29] of some of Boon's work (1964 and 1975), of Bhalla (1964 and 1965), Enos (1974), Timmer (1974), Cooper *et al.* (1975), Kaplinsky (1975), Lal (1975) and El-Karanshawy (1975). The industries for which this conclusion has been reached include salt and sugar production, rice and maize milling, textiles (spinning and weaving) and carpet weaving, brewing and gari processing, the production of metal cans, and metal working industries, shoe production, road building and cement block manufacture. Although the list is large and growing, so that the results can no longer be regarded as isolated incidents, one must note that the finding was probably in part due to the choice of industry. Extensive case studies were, it is likely, only initiated where there was a prior presumption that a range of techniques did exist. The industries listed are generally speaking *old* industries so that old vintages from the advanced countries and/or traditional techniques in poor countries provide labour-intensive alternatives to the latest techniques. It is noteworthy that studies of recently developed industries are much rarer – there is no study, to my knowledge, which includes modern consumer durables or atomic energy. If such were undertaken it is likely that the range of techniques in the core activity would be much smaller. The studies which have been conducted into middle-aged industries, like canning and fertiliser production, have tended to reveal a narrower range of choice, and more tendency for the choice to be confined to relatively investment-intensive operations.

While the studies have established a much wider potential range of choice than that suggested by a technological determinist view, the range established is clearly not infinite, but normally confined to a few discrete alternative techniques, though a 'synthetic' approach, as adopted by some of the studies of the Strathclyde team, greatly increases the number of alternatives. This 'synthetic' approach consists of theoretical marrying of various alternative parts of the production process, thus producing numerous 'combined' alternatives, which are synthetic in the sense that the particular combination of processes postulated may never have been used in a real plant, although the constituent elements have. The development of such synthetic alternatives is itself the product of substantial information collection, and may not be available to decision makers within LDCs without considerable managerial effort. The range of possibilities represented by the discrete alternatives is in some cases itself limited to the investment-intensive end of the spectrum, with even the most labour-intensive alternative representing investment per man considerably above that suggested as appropriate in Chapter 4. For example, the most labour-intensive technique identified for the manufacture of round open-top cans involved investment per worker of £1746 (East African £), which while immensely less investment-using than the most investment-intensive technique used (with $I/L$ of £31,917), was none the less pretty investment-using in relation to the sort of investment per man that Tanzania could afford for its whole workforce – likely to be well below £500.[30] In other cases the labour-intensive or intermediate techniques identified required quite modest amounts of investment per worker – in rice milling, for example, the small rice mill required (fixed) investment per worker of about £330 compared with $I/L$ of over £33,000 for the large bulk mill.[31] Many of the labour-intensive alternatives potentially involved substantially greater employment prospects than the investment-intensive alternatives: for example, in sugar processing in India the most labour-intensive alternative identified uses over three times the labour per unit of output compared with the modern mills (Baron in Bhalla (ed.) (1975)).

Identification of technically efficient labour-intensive techniques as defined here is neither a necessary nor a sufficient condition for such techniques to be the most appropriate choice of technique in a particular environment. It is not a necessary condition because the distributional implications of different techniques may justify the use of apparently inferior techniques; whether the technically efficient labour-intensive technique presents the best choice depends on the relative productivity of the different techniques, as well as economic and social conditions and objectives in the country concerned. A technique may be mildly investment saving in relation to other techniques, but this advantage may be offset – in most economic conditions – by the enormous increase in labour and other requirements. Pickett and Robson's (1976) study of cotton cloth production provides an example.

The studies were, in general, concerned to identify the relative productivity and input requirements of the various techniques, and many show the 'switching-point' factor prices: for example, Timmer (1974), El-Karanshawy (1975), Lal (1975), I.L.O. Kenya (1972), Cooper *et al.* (1975), Baron (1975) and Pack (1975) show required factor prices for the labour-intensive techniques to be ranked first in terms of profitability. While, in the nature of the exercise, a reduction in the relative wage/investment cost price increases the relative profitability of the labour-intensive alternatives, the switching point varies from industry to industry, as do the changes in input requirements and productivity as one switches from one technique to another. In some cases, minor changes in relative prices would bring about significant changes in the choice of technique, assuming techniques are selected according to a profitability criterion; this was the finding of the Strathclyde study on shoes. In others 'efforts to manipulate prices so as to change the decisions made are unlikely to succeed unless the manipulation is massive' (Cooper *et al.* (1975)).

The beginning of this very summary summary used the phrase 'given certain assumptions and within certain limits': it is these assumptions and limits which give support to the more general approach to technological choice outlined above, and described more fully in Chapter 1. The studies showed that the finding of a range of efficient techniques had to be qualified in ways which make more sense of a more general approach, and which suggest that the neo-classical question is often of only minor significance, despite the attention that has been devoted to it. We shall briefly discuss the qualifications below:

*Product quality:* many of the studies showed variations in the nature and quality of the products as between different techniques, and consequently that product considerations could determine technique choice. Thus Baron shows that production of the traditional Indian sweeteners (gur and Khandsari) is both more investment-saving and more employment-creating than either of the two techniques for manufacturing white crystal sugar. The studies in the next chapter shows that the creation of consumer demand for sophisticated maize products gives rise to more capital-intensive grinding techniques; while in block manufacture the demand for products of a certain quality dictates the use of a more sophisticated technology for their manufacture. IBRD (1971) makes a similar point with reference to road construction, and Bhalla and Gaude (1973) (service industries), Kilby (1965) (bread), Gouverneur (1971) (extractive industries), and Pfefferman (1968) with reference to manufacturing in general. In the study by Wells (1973), however, this factor is discounted, it being argued that quality differences in the Indonesian industries investigated were slight and not generally perceived. In his work on metalworking in Mexico, Boon concludes: 'Generally speaking, when physical characteristics are very particularised, the choice of technology narrows down and the co-efficients of production tend to become fixed'.[32] In can making (Cooper *et al.* (1975)) it was noted that the more labour-intensive techniques might reduce the quality of output under certain conditions, and this ruled out the possibility that firms using such techniques 'could supply cans to certain quality demanding segments of the market'. Langdon (1975) found that the capital-intensive techniques operated by multinational companies in the soap industry in Kenya produced a different, 'higher'-quality, more highly packaged and promoted product than the more labour-intensive techniques used by local firms. El-Karanshawy (1975) shows that hand-loom methods produced superior quality carpeting and the mechanised loom products had to be expensively promoted to gain an adequate market.

These differences in products as between techniques emerged in case studies which had been selected, on the whole, to eliminate such differences and compare techniques which produce the 'same' product.

*Economies of scale:* techniques are not perfectly divisible and a definite tendency was found for the more investment-intensive techniques to be designed for a larger scale than the labour-intensive, as a result of the historical stage of development at which the different techniques were developed. Comparisons between techniques were found to be more favourable to the labour-intensive alternatives at small scales of production, while these were often ruled out, becoming inefficient, at large scale. The many extensive studies by Boon support this conclusion. (See also Bhalla and Gaude, and the studies of cement blocks and maize milling that follow.) Scale was also of importance in Baron's study of sugar and Timmer's work on rice-milling.

*Skill requirements:* different techniques required different (in quality and quantity) skills for operation, for maintenance, and for supervision. This conclusion appears in the can-making study and can-sealing studies, the Strathclyde studies, and the case of maize milling, for example. Tests of the Hirschman hypothesis provide indirect evidence of variations in supervisory requirements: broadly, Diaz-Alejandro (1965), Healey (1968), Gouverneur (1971), and Pickett and Robson (1976) support the Hirschman view that the difference between labour productivity in developed and underdeveloped countries is greatest for labour-intensive activities and least for investment-intensive. This is in itself relevant to choice of technique, and also might be interpreted as lending support for the view that labour-intensive techniques may require a greater supervisory effort.

These, and other, variations in characteristics between techniques, in addition to the neo-classical variation of labour and investment requirements, make a vector approach to each technique – with each vector representing an array of characteristics, as described above – preferable to the simple neo-classical characterisation. It then follows that choice among the different techniques will depend on other factors besides the labour/investment price – on, for example, the size of firm and size of market, on the nature of the market and therefore product requirements, on the availability and price of other complementary inputs, such as skilled and supervisory labour, fuel and so on. As argued earlier choice will also depend on the nature of the decision maker because information, access to resources, scale and markets differ between decision makers, as well as the decision makers' objectives.

On the whole, obsession with the neo-classical question has prevented much direct examination of these interrelationships, but they have entered indirectly as explanations of *departures* from neo-classical norms.

Considerable evidence has accumulated of different price of inputs, particularly labour, according to the *scale* and to the *location* (whether rural or urban) of activities. The next chapter provides evidence for maize milling in Kenya. There is evidence for Japan (see Okita (1964)), for India (see Shetty (1963)), and summary evidence for India, Pakistan Egypt and Nigeria by Koji (1966),[33] all suggesting that small-scale enterprises pay substantially lower wages than large-scale. Evidence for differences in the cost and availability of finance has been less systematically collected, but has been widely noted.[34] Prasad (1963) shows that small-scale enterprises pay more for their material inputs than large.

It has been established that the technological choice made by small-scale enterprises tends to be systematically more labour-intensive than large: see, for example, Dhar and Lydall (1961), Sandesara (1966) and NCAER (1972), Staley and Morse (1965), UNIDO (1969), and the sugar, rice milling and maize milling studies. This finding is, of course, related to the finding that larger-scale techniques are likely to be more investment-intensive, to the differences in market (and therefore product quality), and in price of inputs.

Studies comparing the decisions of multinationals (MNCs) and local firms have come up with mixed conclusions: Wells (1973), Langdon (1975), and Morley and Smith (1974) find that MNCs make more investment-intensive choices. Langdon finds this is related to differences in product quality, which Wells does not. Hal Mason (1970) and I.L.O., Kenya (1972) find no systematic differences between MNCs and local firms *if* both are producing the *same* product, but MNCs produce proportionately more investment-intensive products. One would expect product differences to be critical because MNCs are more systematically producing for an international market than local firms and their advantage lies in product differentiation, brand names and heavy promotion. It thus systematically reduces the value of the exercise to confine the comparison to near identical products.

The Morley and Smith study (1974) into MNC decisions in Brazil brings together many of these aspects, showing that their investment-intensive technological choice may be explained in terms of 'satisficing' motivation, scale of production and quality of product. When asked to rank influences over technological choice 17 firms (out of 33) gave product quality as of first significance, 14 gave size of market and only 1 labour costs (though labour costs did appear significantly as a second choice). Generally, firms used the same technology in Brazil as in their home economy – 28 out of 34 firms said there were slight or no differences, other than differences of scale. There was no evidence of any attempt to search for labour-intensive alternatives, which, at least in some cases, would probably have yielded more profitable alternatives. The authors attribute this to satisficing behaviour, and to the relative managerial advantage MNCs have with known and familiar techniques. They show that in about half the industries Brazilian firms used less investment-intensive techniques (measured in terms of electrical energy per worker) than the MNC.

The Morley and Smith study illustrates the significance of the *objectives* of the decision makers: it does not follow, as they show, that a technique which theoretically would maximise profits will in practice be chosen. Wells, Pickett *et al.*, Cooper *et al.*, and Timmer also found examples of decisions which apparently did not maximise profits, and might, according to a narrow definition of rationality, be described as economically irrational. Wells and Pickett attribute these decisions to the predominance of 'engineering man', with his more exacting engineering standards, over 'economic man', who is not worried about engineering standards, only about profits: they may be pointing to a real conflict between decision makers within a large firm, though they need sociological studies of actual decision making to substantiate this. But they might be indicating a conflict within an individual manager arising from mixed motives, which is summed up in a more sophisticated (but possibly also a more obfuscating) way as representing 'satisficing' behaviour by Morley and Smith. It needs pointing out, since the term 'economic irrationality' is often used, that there is nothing irrational about having and effectively pursuing objectives other than profit maximisation. However, more investigation is needed before one can accept the conclusion that the cases studied exhibited non-profit maximising behaviour. Search costs, the cost of alternative technologies, managerial and supervisory requirements of alternatives may have meant that the actual choice was the profit-maximising choice. Moreover, the international context of decision making in the MNC needs to be taken into account. MNCs' advantages lie in rapid and painless access to the latest technology, and consumer adherence to their products, as against local substitutes. While some costs – such as payment for technology and imported parts and machinery – may reduce the on-the-ground profitability, they contribute to the overall profitability of world-wide operations, which a successful search for labour-intensive technology bought from outside the company would not.

The ability of the MNC to exploit and sustain its own oligopolistic advantage brings us closer to what we earlier (Chapter 4) described as the political economy of technological choice. Evidence for the working of the political economy dimension is mainly to be seen in the prevalence of taxes, subsidies, tariffs and exchange rates which promote the use of advanced-country technology;[35] in public sector decisions in favour of the large-scale investment-intensive technique, even sometimes where it is *less* profitable than the alternative (see for example the carpet weaving case); and in the way in which credit and information services operate in favour of the advanced techniques which generate high incomes among the elite. One of the most direct case studies of the political economy of choice is the salt study of Enos. He shows a direct relationship between who gains and how, suggesting a correspondence between the characteristics of techniques, the political power of different groups and the decisions made. He uses the political economy variable to explain the different technical choice made in salt production in three countries in S.E. Asia, in each of which he identifies a different political economy.

APPROPRIATE TECHNOLOGY

Chapter 4 discussed and defined the characteristics of appropriate technology: as argued there appropriate technology has a number of characteristics of which its relative labour-intensity is one, but only one. Others include small scale, simple products, provision for the spread of income-earning opportunities, and so on. Thus the case studies just considered are not, for the most part, concerned with the existence and viability of appropriate technology considering *all* its characteristics, but only with the question of the existence and efficiency of more labour-intensive techniques. Consequently, the light they shed on the appropriate technology question is limited and indirect – indeed many of the studies exclude alternatives involving different and more appropriate products altogether, as argued above.

None the less, some of the studies did identify efficient techniques which would qualify as being appropriate in many respects. This was true of Kaplinsky's study of gari processing, the studies that follow of cement block manufacture and maize milling, Baron's sugar study, Timmer on rice milling and El-Karanshawy on hand-loom carpet manufacture. Thus to the limited extent that the studies do consider techniques with broadly appropriate characteristics, they find many of them efficient in many economic conditions (although it must be noted that hand methods of cotton-weaving and the Ambar Charkha for cotton spinning were found to be of relatively low efficiency).[36] Moreover, in so far as the term appropriate is used of any labour-intensive techniques then, as shown above, the studies identified many examples which would, in certain circumstances, prove the most sensible choice of technique.

Work by the Intermediate Technology Development Group and others concerned to promote the cause of intermediate and appropriate technology has been action-oriented – rightly in view of the general inaction in this area. These groups have identified many techniques which they define as appropriate:[37] the range is wide, as a cursory glance at their work suggests – it includes a Chinese chain-lift pump, a metal-bending machine, a paper-pulp packaging unit, a foundry unit, hospital equipment, a cassava grinder, a bamboo fruit picker, house-building techniques, a quack stick, rainwater catchment tanks, a rice transplanter platform, methane as a source of energy, solar water energy, leaf protein, kitchen redesign; and so on, and so on. Many of the techniques involve radically different products and scale from those produced by formal advanced-country technology. This is one reason why they have not, on the whole, been subject to full-scale economic case studies. Even if they were the results would critically depend on the weighting given to income distribution factors, since in general (although not always) they involve a completely different distribution of earning opportunities from the advanced technology alternative. These techniques, together with some of the economic case studies discussed above, establish the existence of viable appropriate techniques in a number of fields. The existence of *some* techniques is not in doubt; what is not yet established is the extent to which alternatives to advanced technology exist across the board, permitting a wholesale switch from advanced technology.

The process of development and discovery is of course an on-going one. Work on appropriate techniques has established the possibility of creating alternative techniques, given resources and determination. For example, whiteware pottery, brickmaking, Portland cement production, and egg packaging production have been redesigned for use on a small scale, while the International Rice Research Institute has been responsible for many improved designs of farm equipment and tools.

While these examples establish that technological developments can be redirected in more appropriate directions, it remains true that the bulk of the world research effort continues to be devoted to the search for more inappropriate techniques, making both the techniques and the economic case studies focused on them rapidly out-dated.

NOTES

1. The two bulletins of the ITDG, first *IT Bulletin* (1967–73), then *Appropriate Technology* (1974 to date), are much the best source of information on this type of technology, presenting many fascinating examples, which will be further discussed later in the chapter. A bibliography of such cases has been compiled by Marilyn Carr (1976).
2. For a full bibliography of such case studies up to 1974, see Jenkins (1975), and my introduction in Jenkins.
3. The attack on the concept was initiated by Robinson (1953–4). Harcourt (1974) provides an excellent easy guide to the debate. See also Sen (1974) for an amusing and enlightening review.
4. In general throughout this book some attempt has been made to use the terms *investment-intensity*, *investment-saving*, etc., rather than *capital-intensity*, *capital-saving*, but occasionally the old terms seem to have crept back in unnoticed.
5. The view that some 'correction' is required, and the nature of this correction, form the heart of recent manuals on social cost/benefit analysis and project selection in LDCs – see especially Little and Mirrlees (1969), and the UNIDO Guidelines (Das Gupta *et al.* (1972)). Discussion and criticism of these approaches are contained in a special issue of the *Oxford Institute of Economics and Statistics Bulletin* (February 1972) and in Stewart (1975).
6. This is, for example, the method used in subsequent chapters.
7. Boon's early studies (1964) use prices obtaining in the Netherlands. A good deal of the Strathclyde work seems similarly to be based on prices in the supplying country.
8. Attempts to 'correct' prices are made, for example, by Lal (1977).
9. Sen (1968), pp. 101–4 uses the formula

$$w_k = (m + l \, . \, w)n \tag{1}$$

where $w_k$ = working capital requirements per unit of output;
$m$ = material and fuel costs per unit of output;

$n$ = average time lag between incurring costs and sales, expressed as a proportion of the period to which costs relate;
$l$ = labour requirements per unit of output;
$w$ = wage rate;

$$W_k = w_k \, . \, O,$$

where $W_k$ = total working capital requirements,
$O$ = level of output,

and

$$O = \frac{I}{v}$$

where $I$ = fixed investment;
$v$ = investment output ratio.

So,

$$w_k = \frac{w \, . \, k \, . \, I}{v}$$

and

$$\frac{W_k}{I} = \frac{wk}{v} = \frac{(m + l \, . \, w)}{v} \, . \, n \qquad (2)$$

(2) above shows the ratio of working capital requirements to fixed investment. Since $v$ is normally higher for (fixed) investment-intensive techniques and labour requirements ($l$) are lower, it is likely that the ratio of working capital requirements to fixed investment will be higher for labour-intensive techniques than for fixed-investment-intensive. However, there might be some offsetting tendency arising from lower wages ($w$) paid by labour-intensive techniques, but then differences in $m$ (probably higher for labour-intensive techniques) and $n$ (possibly lower) also need to be taken into account.

10. Sen (1968, Appendix C) showed how the inclusion of working capital could alter the ranking of techniques in terms of surplus generated per unit of investment. The hand-loom techniques gave greatest surplus where working capital was excluded, but when it was included the ranking was reversed with the power looms maximising the rate of surplus.
11. The present value, $P$, of a constant infinite series of annual payments, $A$, is given by the formula,

$$P = \frac{A}{r}$$

Thus to express working capital as an annual *cost*, using this formula, *gives*

$$\overline{W}_k = r \times W_k$$

where $\overline{W}_k$ is the annual equivalent of a total working capital requirement, $W_k$, showing that the burden of working capital varies with the interest rate. Similarly (as shown in subsequent chapters) the annual burden of fixed capital varies with the interest rate. With zero interest rate, an annual equivalent cost remains for fixed investment, but working capital becomes costless.

12. E.g. Boon (1964).
13. Lancaster's proposed method of valuing different characteristics of products is based on the use of market prices; the 'hedonic' indices are similarly based.
14. Little and Mirrlees (1969).
15. See I.L.O. (1972), technical paper 7.
16. See Phelps (1963).
17. Cooper and Kaplinsky (1974) make this point and illustrate it empirically.
18. The distinction is also between product and process industries – Gouverneur (1971), Chapter IX.
19. The key position of scale has also been established by studies by Boon. Cooper's

study of can sealing (I.L.O., 1972) shows that the less automated method becomes the least costly method at low rates of capacity utilisation.

20. Suppose total costs, $TC = w \,.\, l \,.\, \bar{O} + Ac$, for the investment-intensive technique, and

$$TC' = w \,.\, l' \,.\, \bar{O} + Ac'$$

for the labour-intensive technique,
where $w$ is wage rate;
$l$ labour input per unit of output;
$\bar{O}$ full capacity output;
$Ac$ annual investment costs, for investment-intensive technique;
and $l'$, $Ac'$ for labour-intensive.

By assumption full capacity output $\bar{O}$ is the same for both techniques (this might be because they were of the same size, or because there were more machines in use for the labour-intensive technique). Labour is assumed to vary directly with output, so all labour is variable, none overhead. Assuming that capacity utilisation falls to $\alpha\bar{O}$,

Then the new $T.C. = w \,.\, l \,.\, \alpha\bar{O} + Ac$
$T.C.' = w \,.\, l' \,.\, \alpha\bar{O} + Ac'$

and the decline in total costs is $(l - \alpha)w \,.\, l \,.\, \bar{O}$ and $(l - \alpha)w \,.\, l' \,.\, \bar{O}$. By assumption, at full capacity $T.C.' = T.C.$, and $Ac/l > Ac'/l'$. Hence $l < l'$, and with the same wage rate $w \,.\, l \,. < w \,.\, l'$. It follows that the decline in total costs, with decreasing capacity utilisation, is greater for labour-intensive than for investment-intensive techniques.

21. The factor emphasised by Winston as explaining rates of capacity utilisation.
22. 'Due to lack of working capital and the very high rate of interest that has to be paid, production in hand industries is low and unsteady, and cost of production is higher' (Prasad, p. 160). Interest rates paid by hand industries he found were 16 per cent compared with 8 per cent paid by mill industries.
23. By Pack (1974).
24. Example suggested by R. Jolly.
25. Sen, in his introduction to the third edition, presents a formula giving some weight to consumption and some to surplus generation. This is the same sort of approach as that adopted by Little and Mirrlees with their premium on savings, which still gives some weight to consumption. Galenson-Leibenstein (1955), Dobb (1956–7) and Sen, first edition, all appear to give overriding weight to savings generation.
26. See Sen (1968) and Bhalla (1964b).
27. See Okita (1964), Shetty (1963), Koji (1966).
28. The concept of technical efficiency was defined (and criticised) in Chapter 1. To recapitulate: techniques are described as technically efficient if for any given output level, in relation to each other, the use of more of one factor is associated with the use of less of some other factor. A technique is said to be technically inefficient or inferior if it uses more of each factor to produce a given output. Technical efficiency is generally a necessary but not a sufficient condition for economic efficiency. An economically efficient technique is one which minimises cost – private or social depending on the point of view adopted. Economic efficiency thus depends on technical factors and on economic variables, particularly prices, though as argued above one cannot in fact define the technical factors independently of economic variables. Note that this definition has nothing to do with engineers' definition of technical efficiency.
29. Preliminary results of two of these studies appeared in Pickett *et al.* (1974); these two were published in full in Strathclyde (1975). Further investigations have appeared in mimeographed papers and will be published in *World Development* and by HMSO.

30. See Cooper *et al.* in Bhalla (ed.) (1975).
31. Timmer (1974).
32. Boon in Bhalla (ed.) (1975).
33. Koji summarises some of the evidence as shown below:

Wage Index according to size of establishment

<table>
<tr><th>No. employed</th><th>India (1955)</th><th>Nigeria</th><th>Pakistan</th><th>U.A.R. (1961)</th></tr>
<tr><td>5–9</td><td>—</td><td>—</td><td>58</td><td>—</td></tr>
<tr><td>10–19</td><td>47</td><td>—</td><td>76</td><td>—</td></tr>
<tr><td>20–49</td><td>51</td><td>75</td><td>89</td><td>—</td></tr>
<tr><td>50–99</td><td>55</td><td>83</td><td>96</td><td>65</td></tr>
<tr><td>100–199</td><td>72</td><td>67</td><td rowspan="4">100</td><td rowspan="2">74</td></tr>
<tr><td>200–499</td><td>85</td><td rowspan="2">73</td></tr>
<tr><td>500–999</td><td>88</td><td rowspan="2">100</td></tr>
<tr><td>1000+</td><td>100</td><td>100</td></tr>
</table>

SOURCE Koji (1966), Table 1-B.

34. See for example Shetty (1963), Staley and Morse (1965), Prasad (1963), Todaro (1969), UNIDO (1969), McKinnon (1973), Chandavarkar (1970). McKinnon finds that in Ethiopia small-scale manufacturers (largely rural) pay as much as 100 to 200 per cent for finance, while larger urban firms pay 8–9 per cent p.a.
35. The macro studies of particular countries illustrate this – e.g. Little, Scitovsky and Scott (1970); and Strassman (1968).
36. See Raj (1959), Sen (1968), Bhalla (1964(b)).
37. See Marsden (1971), G. McRobie and M. Carr (1976), the references in that article, the Carr bibliography, and the journals of the ITDG.

# 9 The Choice of Technique — Maize Grinding in Kenya

Maize is the staple food in large parts of Kenya.[1] In 1966 it was estimated that on average 13–17 million bags, or 250–300 lb per head were produced each year.[2] It is eaten in a number of ways, but is most often consumed in the form of a flour. In this form it is probably eaten by the majority of Kenyans at least once a day. Hence the process of grinding maize is potentially very big business affecting almost everyone. Maize is a subsistence crop and in normal conditions about 80 per cent of total production is not marketed but consumed by those who produce it.[3] There is a wide range of methods of processing maize in use in Kenya. They extend to traditional methods, which pre-date historical records, involving the use of a wooden pestle and mortar, to the most modern roller mills, of identical design to those currently in use in the developed countries. Between these two extremes are hand-operated mills, water and grindstone mills, and hammer mills which are operated either by diesel engine or electric motor.

The industry thus offers an example of a profile of choice of techniques varying from the ancient to the modern, in scale and location of operation, in investment- and labour-intensity, in quality of product and source of power. A survey of the techniques in use was therefore carried out in Kenya in 1969 in the hope that this might help illuminate choice of technique in maize grinding, and also more generally.

The Survey covered 88 firms, together responsible for 103 different machines. The areas covered in the Survey[4] are shown in Table 9.1.

No attempt was made to make the sample representative of Kenya as a whole: the aim was to get a sufficiently large number of examples of each type of technique to be able to generalise about the operation of each type.

## TYPES OF TECHNIQUES

1. *Traditional methods.* The only 'machinery' involved in these methods is the use of a wooden stick and bowl, made domestically or by the local carpenter. The methods have been fully described by Dr Schlage:[5]

The amount of maize required for several meals is taken off the cob and transferred to a wooden mortar. One or two cups of water are added and the whole amount is pounded with a wooden stick stamped into the mortar for about 15 minutes. Afterwards the maize is transferred into a flat basket and by shaking the grains are separated from the peels. The grains are again collected into the mortar and stamped in the same way as before for three or four shorter periods followed by the separation of the peels. The resulting product is called *pure*. The bran is quite often used to feed chickens. Some of this pure is cooked as it is or together with beans. Most of the pure undergoes a further processing starting with soaking in excess water for three or four days until a smell starts due to fermentation. The water is poured off, the soaked pure washed about three times with fresh water, decanted, and again pounded in the mortar. After twenty minutes the pure is separated as before from the finer parts or even sieved off. The bigger particles are kept for further pounding until everything is completed to a semolina-like flour. This product is ready for use for preparing *uji* or *ugali*. It cannot be kept for more than a day except after drying in the sun which allows storage for some days.[6]

These, and similar methods, are purely domestic and subsistence activities, not carried out for exchange. The capital investment involved consists entirely of local labour and minimal local materials. Thus the methods can be thought of as almost 100 per cent labour-intensive.[7] These methods do not represent a real alternative to the other methods considered, since they are never practised commercially, and even domestically they are rapidly giving way to the other methods. Hence they were not included in the survey. Their nutritional implications are considered, with the others, below.

TABLE 9.1

| *Province* | *No. of firms* |
|---|---|
| Central | 21 |
| Eastern | 31 |
| Nyanza | 9 |
| Rift Valley | 8 |
| Western | 14 |
| Nairobi | 5 |
| Large towns (over 10,000) | 15 |
| Small towns (1000–10,000) | 9 |
| Rural | 64 |

2. *Hand-operated mills.* These small mills are operated by turning a handle: they are very similar to hand-operated coffee grinding machines or mincing machines. Generally speaking they are used for preparation of domestically grown maize for domestic own-consumption, though they are lent to neighbours and maize is sometimes ground with them in exchange for supplies of maize or even cash. However, in all the cases covered in the survey the hand mills were primarily devoted to home-use.

3. *Water mills.* These mills were the first non-manually operated methods to be adopted. Water provides the power which turns propellors, which turn stones of between two and three feet in diameter. The stones grind the maize. Very little labour is required as the mill can operate without supervision and labour is only needed to pack, weigh, etc. Some of the stones covered in the survey were imported from India but many are now made locally.

4. *Hammer mills.* These mills vary in size. They are powered with electric or diesel engines. Originally imported from the U.K., U.S. and Germany, they are now made locally in three sizes. They grind all the maize to a brownish flour, known as posho, or ordinary posho (to distinguish it from sifted, see below). The roughest parts of the shells of the maize are sometimes discarded, and sold as cattle-feed, but most of the maize is retained. In particular the majority of the germ and bran is retained, which is of importance nutritionally, though it also means that the flour does not keep very long, normally for about two or three days. Labour requirements are for preparing the whole maize, weighing and packing the maize and flour and supervising the machine.

5. *Roller mills.* These involve a quite different process from the hammer mills. A series of machines is involved which clean, grind, sift and discard. The maize is transferred automatically – by chute – from one stage to the next. Labour requirements are for loading the machine, supervising, maintaining and repairing the machines and packing the finished product. A modern roller mill takes up a good deal of space: visually it makes a considerable impact as the vast area of pipes and machines is almost completely empty of mankind, in sharp contrast to the packing room which is full of people. The roller mills in Kenya are all imported – from Germany mainly but also from the U.S. and U.K. Roller mills were introduced into Kenya relatively recently. In 1935 the first roller mill was established at the Coast to process maize imported from South Africa. The sifting process involves removal of the bran and germ. This affects the nutritional content, the taste and texture and the keeping qualities – it keeps considerably longer than ordinary posho. The bran and germ which is removed is generally used for cattle-feed. Roller mills produce three types of flour: sifted flour grade I; sifted grade II (less white and finely ground); and granulated (unsifted), which is similar to ordinary posho produced by the hammer mills.

TABLE 9.2 Nutritional aspects of different types of maize

| | Moisture % | Calories per 100 gm | Protein % | Fat % | Carbohydrates % | Fibre % | Ash % | Calcium | Iron | Thianine | Niacin |
|---|---|---|---|---|---|---|---|---|---|---|---|
| | | | | | | | | mg per 100 gm | | | |
| 1. Whole maize | 12 | 355 | 9·2 | 3·9 | 73·7 | 1·6 | 1·2 | 10 | 2·4 | 0·38 | 2·0 |
| 2. Whole maize* | 13·5 | 347 | 8·5 | 3·8 | 70 | 2·7 | 1·2 | | not | available | |
| 3. Trad. methods* | | | | | | | | | | | |
| Pure | 16·2 | 335 | 7·4 | 1·1 | 74 | 1·0 | 0·4 | | not | available | |
| Unga | 35·7 | 259 | 5·7 | 0·9 | 57 | 0·5 | 0·2 | | not | available | |
| 4. Hammer/water mills† | 12·2 | 353 | 9·3 | 3·8 | 73·4 | 1·9 | 1·3 | 17 | 4·2 | 0·30 | 1·8 |
| 5. 'Degermed', i.e. sifted flour, grade I‡ | 12 | 363 | 7·9 | 1·2 | 78·4 | 0·6 | 0·5 | 6 | 1·1 | 0·14 | 1·0 |

* Schlage (1968), especially Table 2.
† F.A.O., *Food Composition Tables for Africa 1968*.
‡ *Maize and Maize Diets*, F.A.O., No. 9, 1953.

The products of the different methods are thus not homogeneous. They differ in nutritional content, taste, and, partly as a consequence, in markets.

NUTRITION

Oddly, the oldest and the newest methods have much the same nutritional implications: both remove a good deal of the germ and bran and with this some of the important sources of nutrition. Table 9.2 shows how the sifting process (traditional and modern) removes protein and important minerals from the maize. The reduction of vitamin B, niacin, is particularly important as maize contains an element antagonistic to this vitamin and consequently people who eat a lot of maize need more of it than others. This vitamin is important for mental alertness. The absolute reduction in protein does not tell the whole story, as of the three types of protein contained in maize – zein, globulines and gluteline – the better types (globulines and glutelines) are removed, leaving zein, which is a poorer source of protein. Essential fatty acids, oleic and linoleic, are also removed with the removal of the germ. Some of the firms producing sifted flour also produced 'fortified' flour with added vitamins and protein. These products do not contain as many nutritional

TABLE 9.3 Output per bag (200 lb) of whole maize

| *Method* | *Rejected altogether or used for cattle/poultry feed (lb)* | | *Flour for human consumption (average) (lb)* |
|---|---|---|---|
| | *Range* | *Average* | |
| Traditional methods* | 89 | | 111 |
| Hand crusher | 4–50 | 28 | 178 |
| Water mills | 1–12 | 6·3 | 194 |
| Hammer mills | 0–7 | 0·9 | 199 |
| Roller, sifted II | 5–10 | 8·3 | 192 |
| Roller, sifted I | 20–66 | 43·6 | 156 |

SOURCE

* from Schlage (1968) p. 13.

Remainder from survey. They are in line with figures quoted in evidence to the Maize Commission of Inquiry: a gristing loss of 1–2% was claimed for ordinary posho from a hammer mill and one of between 20 and 40 lb for sifted products. (Evidence of Walker, an associate agent of the Maize and Produce Board, M. H. Patel, a miller, the General Manager of Unga, one of the largest mills, and Damji, Shah, Chandaria, C. J. Patel, all millers.)

elements as the original maize, or hammer mill products, cost more than the unfortified sifted flours, and are not widely consumed.

The figures in Table 9.2 show the percentage of nutritional elements for a given amount of maize or maize product. But there are also considerable variations in grinding loss which means that a given quantity of whole maize results in different quantities of flour according to the method adopted. This arises because the sifting methods (traditional and modern roller) involve the rejection of germ and bran and consequently a reduction in the ratio of final product to input, as Table 9.3 shows.

### TASTES AND PREFERENCES

Sifted maize meal is a relatively new product. It was originally imported from S. Africa for consumption in Mombasa. The first roller mill (a small one) was set up at the Coast in 1935. The Nairobi mills were only established on any scale in 1955. According to the general manager of one of the largest producers of sifted maize: 'In a normal market we would try to produce what the market required. *The sifted maize meal was something my Company pioneered in this country, and it has created a market for itself.*'[8] This statement, and the history of maize consumption in Kenya, suggests the inherent contradiction between taking consumers' preferences as the ultimate guide to production decisions and welfare, and spending money on determining these tastes. The managing director of another major producer of sifted flour explained that even when the sifted flour was not available people would not eat posho because 'we [i.e. producers of sifted] have considerable advertising and pains and personal contacts and what-not to pick up a good market.'[9] The producers of sifted flour do spend considerable sums on promotion. Evidence to one inquiry[10] suggested that selling expenses could add as much as one-third to production costs.

Although the producers created the market for the sifted products the preference for these products, particularly in the urban areas, is genuine. The sifted flour tastes sweeter, lasts longer and tends to be cleaner than ordinary posho. Evidence from different witnesses to the Maize Commission explains the preferences:

> But the people in the urban areas were not interested in eating this impure food [i.e. ordinary posho] and in the time of civilisation and what-not they decided they were interested in eating something better, the purer maize meal. (A director of a company producing sifted products)[11]

> They [the average African] prefer the sifted flour, they are liked by everybody, but when you go out, to the Masai for example or in Kitui or in remote parts then any maize meal you produce you find a market for. (Ngure)[12]

> It's [ordinary posho] kept in a dirty way, sometimes you buy it you find that the posho itself, inside, is having some dirt, and sometimes you find it has got a bitter taste. Sometimes it changes – sometimes a bitter taste, sometimes a different smell. (Onguso, storekeeper for the C.I.D. H.Q.)[13]

It was agreed that the demand for sifted flour was concentrated in the urban areas. According to one witness, 'only the educated people in Offices [eat sifted], but not in the reserves' (a miller, producing ordinary posho).[14]

> Posho is more popular with more people. That [i.e. sifted] is for people who can afford it but in the country life let's go in for something which is normal. (Senator Machio)[15]

One of the witnesses who preferred sifted flour himself agreed that ordinary posho was fresher in the rural areas where people took their own maize to the mill and waited for it to be ground: 'If it [ordinary posho] is in a clean way I think that's what I was brought up on. Sometimes I used to take it myself. My mother used to send me, and it used to be milled in my presence and it was so nice. Very nice.' (Onguso, storekeeper at C.I.D. H.Q.)

Well over three-quarters of the maize produced in Kenya is consumed by the producers. It is eaten whole, ground domestically or taken to a local hammer or water mill. Thus the majority of maize consumed is not sifted. But as people become urbanised, they stop consuming maize they themselves have grown and tend to switch to sifted products. In Mombasa, which is not in a maize-growing area, virtually all the flour is sifted. In Nairobi a rapidly increasing proportion of flour consumed is sifted. The trends were indicated, in the survey, by the high and increasing demand reported by the large roller mills, many of which were working a 24-hour day. In contrast hammer mills in and around Nairobi were being operated at most on a one-shift basis, and some less. Some had closed down altogether, and all attributed this to the growing popularity of sifted products.

### MARKETS

The different techniques thus cater for different segments of the market. Traditional processing methods and the hand mills are almost entirely domestic activities in which members of the family prepare the maize they have grown themselves for their own consumption. Hand mills are occasionally lent to other families and sometimes their maize is ground on the mills in return for some other service, or even occasionally cash. But generally the use of hand mills is not a commercial activity. The water and hammer mills in the rural areas and small towns are largely used – on a commercial basis, i.e. in return for cash –

to grind maize for local farmers. Almost all the mills covered catered for customers within a fairly small area – two or three miles at most. The larger hammer mills receive whole maize from the Maize and Produce Board and sell it to licensed retailers. The roller mills buy maize entirely from the Maize and Produce Board: their products are packed in 5 lb paper bags, branded and distributed throughout the country. It is worth emphasising the difference in markets – and trends in demand – for the products of the different techniques – because in this case choice of technique, or method of processing maize, involves choice of the nature of the product. Consequently one cannot draw any conclusions from the nature of production methods involved without also looking at the implications for consumption patterns.

### SURVEY DATA

Hammer mills accounted for over 70 per cent of the machines included in the survey, roller mills for 12½ per cent, water mills for nearly 12 per cent and hand mills for 5 per cent, as shown in Table 9.4. These figures

TABLE 9.4 Number of examples of techniques

| *Techniques* | *Rural* | *Small towns* | *Large towns* | *Total* | *% of total* |
|---|---|---|---|---|---|
| Hand | 5 | – | — | 5 | 4·9 |
| Water | 11 | 1 | — | 12 | 11·7 |
| Hammer | 53 | 8 | 12 | 73 | 70·9 |
| Roller | 6 | 1 | 12 | 13 | 12·6 |

are not representative of the distribution of different methods in Kenya as a whole: in particular hand mills are certainly underrepresented in the sample since no systematic attempt was made to identify more than a few examples of hand mills. Roller mills are overrepresented, since all the major roller mills outside Mombasa were included, whereas only a proportionately small sample of hammer mills was included. Table 9.4 confirms the rural/urban distribution of the techniques discussed above: the hand mills were only found in the rural areas: the water mills were virtually confined to the rural areas with one example in a small town. Hammer mills were to be found in all areas, but predominantly in the rural areas. The roller mills were almost entirely concentrated in the large towns.

The information received about each of the techniques varied considerably between different users. This is indicated in the tables below by the figures for range, standard deviation and coefficient of variation. The variation shown is not surprising: indeed it is to be expected, given the many differences between the examples in, for example, size and origin of machine, size, quality and purpose of buildings, date of purchase, capacity utilisation and quality of management. The variation makes it difficult to generalise about the techniques, and thus to draw firm conclusions. However, as will be indicated in more detail below, although there were substantial variations within each category, generally speaking the differences between categories were greater. The

TABLE 9.5 Hammer mills: average purchase price (E.A.*s*) (number observed in brackets)

| | *Local* | | *Imported—small* | | *Imported—large* | |
|---|---|---|---|---|---|---|
| DIESEL | | | | | | |
| Mill only | | | | | | |
| new | 3000 | (12) | 3300 | (16) | 20,000 | (1) |
| second-hand | 2000 | (1) | 2600 | (7) | — | |
| Total | 2900 | (13) | 3100 | (23) | 20,000 | (1) |
| Mill plus engine | | | | | | |
| new | 10,200 | (20) | 11,900 | (17) | 70,000 | (1) |
| second-hand | 6800 | (1) | 8300 | (8) | — | |
| Total | 10,000 | (21) | 10,800 | (8) | 70,000 | (1) |
| ELECTRIC | | | | | | |
| new | 4100 | (4) | 7200 | (5) | 30,000 | (1) |
| second-hand | — | | 3000 | (1) | 20,000 | (1) |
| Total | 4100 | (4) | 6500 | (6) | 25,000 | (2) |
| BUILDINGS | 6300 | (17) | 15,600† | (20) | 80,000 | (1) |
| | | | 34,000* | (23) | | |

* Including three relatively expensive buildings, which include stores as well as housing the mills.

† Excluding these three expensive buildings.

variations should, thus, make one cautious about *precise* comparisons – which should be borne in mind for the rest of the discussion, which, perhaps, does not sufficiently qualify the comparisons made, and the conclusions reached. But they need not rule out comparisons entirely. Most studies[16] of choice of technique take a single set of figures to represent each technique, and thus appear to avoid the problems encountered here. This simply means that normally the problem of variance has been assumed away *ab initio*, not that it does not, potentially, exist.

Some of the differences are undoubtedly due to differences in the accuracy of the information provided. The water and hand mills were fairly homogeneous in character. The roller mills were all different in almost every respect – design, manufacturer, date of origin, capacity, etc. There were a number of sub-categories within the hammer mills: first, there was a distinction between those powered electrically and those which used diesel engines; secondly, between locally produced mills and imported mills (all engines were imported); thirdly, there were two examples of imported mills that were substantially larger than all the rest, and therefore need separate treatment; fourthly, there was a distinction between those bought new and those bought second-hand. These distinctions are brought into the calculations below, where relevant.

*Purchase price*

1. *Hand mills:* the five hand mills cost between 160/- and 300/-, with an average of 206/-. The most expensive was bought most recently, and was the only example of a locally produced hand mill. 300/- is therefore the best guide for a new mill. Hand mills do not, of course, need their own buildings.
2. *Water mills:* the average cost of grinding stones was 630/-. (This covered, on average, two stones.) The replacement cost of a stone in 1969 was 350/-. There were substantial variations in the cost of buildings, which varied from as little as 200/- to 20,000/- in one case (which included the building of a dam). Excluding this exceptional case the average cost of buildings was 1730/-; the average cost of stones plus building was 2360/- excluding the dam case, and 4390/- including it.
3. *Hammer mills:* Table 9.5 shows the average cost of different sub-categories of hammer mills.

As Table 9.5 shows, once one excludes the two large imported hammer mills, there is very little difference in purchase price between the imported and locally produced mills: both cost on average around 3000/- alone and around 11,000/-, including engines, when bought new. Mills bought second-hand were only a bit cheaper than mills bought new, though second-hand engines were substantially cheaper. To some extent these comparisons may be misleading since the average age (and the date when bought) varied between the different categories. The local mills were much younger than the imported ones: on average three years old, compared with ten years for the comparable imported mills. However, this is of greater relevance when we come to discuss repair costs than purchase price, though there undoubtedly has been some price rise in mills over the period: thus the imported mill price understates today's price somewhat more than the local mill price.

A substantial difference between the local and second-hand mills occurs in the cost of buildings: on average, buildings associated with the imported mills were 2·5 times (excluding the large stores) or 5·4 times (including the large stores) as expensive as those associated with the local mills. This difference does not arise out of different technical requirements: the mills are technically virtually identical. It arises out of differences in the type of business associated with the different machines. Buildings may be more or less elaborate according to their quality, their size, whether they include sleeping accommodation, large or small storage accommodation, etc. In general the imported mills (which were bought earlier historically) tend to be associated with a grander type of business. They were more often Asian-owned, whereas the later local mills are almost entirely African-owned. This may explain a grander style of building since the Asian entrepreneurs had more capital with which to build. For various reasons capacity utilisation also seems to have been higher some years ago: this meant that the business could be expected to be more profitable, justifying greater expenditure on buildings and also requiring more storage space. The difference in building costs (which outweighs any differences in machine costs) illustrates the dangers of believing that one can associate a particular technique with a unique 'quantity of investment'.

Generally, electrically-operated mills are cheaper to install than diesel-powered mills, but this is only an alternative where electricity is available. There were too few examples of the larger hammer mills to make any generalisation possible: statistics for them are only included in some of what follows.

4. *Roller mills:* these varied enormously in price – from 28,500/- for plant and equipment to 4·7 million shillings. The average price for each of the thirteen separate mills was 594,500/-. The associated buildings varied from 60,000/- to 1·7m (which also housed a wheat mill).

Despite the many variations the figures indicate clearly that the initial capital cost of the techniques is much greater for roller mills than hammer, for hammer than water and water than hand mills. Table 9.6 summarises the position.

The hand mill provided employment for one person at a time – though

TABLE 9.6 Initial cost (E.A.*s*., average)

| | *Plant and equipment* | *Buildings* | *Total* |
|---|---|---|---|
| 1. Hand crusher | 210 | — | 210 |
| 2. Water | 630 | 1730¶<br>3760‖ | 2360¶<br>4390‖ |
| 3a. Hammer —local* | 10,000 | 6300 | 16,300 |
| 3b. Hammer —imported (small)* | 10,800 | 15,600‡<br>34,000§ | 26,400‡<br>44,800§ |
| 3c. Hammer —imported (large)† | 40,000 | 80,000 | 120,000 |
| 4. Roller | 594,500 | 243,600 | 838,100 |

* Excludes electrically operated
† Includes electrically operated
‡ Excluding 3 stores
§ Including 3 stores
¶ Excluding large expenditure for dam
‖ Including large expenditure for dam

NOTE This is average building cost *per mill*. Where buildings house more than one mill the figures have, therefore, been adjusted downwards.

unpaid and discontinuous employment. It was part of the subsistence economy. Employment on the water mills varied from one to three (if we include the owner), with an average of $1\frac{1}{2}$. The water mills operate very slowly, sometimes taking a whole day to grind one bag of maize, and while they operate they require little supervision. Hence they generate very little employment for the activities associated with mills – weighing, packing and supervising. They also involve little repair and maintenance (see below) and require little employment for this.

The small hammer mill employed between one and four people, typically two or three. Since the mill's throughput was greater than the water mills' (three or four bags an hour typically), more employment was required for weighing, loading, packing, etc. The diesel engines required more supervision than the water mills. The large imported hammer mill used little more labour than the smaller ones – on average three compared with two on the small mills.

Employment generated by the roller mills varied according to the size of the mill, from four to 116 per shift. In the large modern roller mills very little labour is used for the operation of the mill, which is almost entirely automatic, and simply needs maize to be put in at one end and taken out at the other. The employment generated is of two varieties: maintenance and supervision, and packing. The largest mill employed 35 people a shift on the mill itself, of whom 19 were skilled, and a further 245 (who covered all three shifts) for packing, thus illustrating the proposition that investment-intensive techniques create employment in their ancillary activities. Table 9.7 shows the average, highest and lowest, investment cost per employee, assuming one-shift operation in each case.

A roller mill costs 4000 times as much as a hand mill, and nearly 200 times as much per worker employed. These figures mean that for any given capital outlay a hand mill would provide some employment for

TABLE 9.7 Initial investment cost per worker[h] (E.A.*s.*: one-shift operation)

| | *Plant, equipment and buildings* | | | |
|---|---|---|---|---|
| | *Average*[a] | *Highest* | *Lowest* | *Standard deviation*[b] |
| 1. Hand | 210 | 300 | 160 | 65 (31·5%) |
| 2. Water | 1960[c] 3570[d] | 20,800 | 220 | 1290[c] (65·8%) |
| 3a. Hammer: Local mill | 8380 | 16,400 | 3670 | 3160 (43·1%) |
| 3b. Imported (small) | 8350[e] 12,830[f] | 52,000[g] | 1750 | 5790[e] (69·3%) |
| 4. Roller | 41,180 | 82,000 | 14,300 | 19,200 (46·6%) |

[a] In this, and all subsequent, tables, the average is based on the data for all cases for which relevant information is available. This means that the average does not always cover exactly the same population, which may lead to apparent inconsistencies between the tables.
[b] The bracketed figures show the coefficient of variation.
[c] Excluding mill involving exceptional dam expenditure.
[d] Including mill involving exceptional dam expenditure.
[e] Excluding three exceptionally expensive buildings.
[f] Including three exceptionally expensive buildings.
This notation will be used in subsequent tables without further explanation.
[g] Includes very expensive buildings, and is also a case where employment has declined substantially from peak period, for which it was designed.
[h] Working owners, or managers, have been included as workers.

200 times as many people as a roller mill – indeed more, given that the hand mills are passed around from person to person. The same expenditure on water mills would provide nine times as much employment as if it were devoted to roller mills; on hammer mills (local) six times as much employment, and imported (small) mills 3·6 times as much employment.

*Investment costs per unit of output*

To start with, to avoid questions of capacity utilisation and variations in product quality, output is defined as the number of bags ground per hour.

The output per hour of the hand mills depends on the energy put into milling, and varies from person to person. The three estimates received were for $\frac{1}{25}$, $\frac{1}{20}$ and $\frac{1}{8}$ of a bag per hour. The water mills also showed considerable variation – this time depending partly on the force of the water flow. The lowest estimate was for $\frac{1}{14}$ of a bag in an hour; the highest for one bag per hour. The typical speed seemed to be between $\frac{1}{4}$ and $\frac{1}{3}$ of a bag an hour. There was no systematic difference in bags per hour between the local and imported mills; rates varied according to the capacity of the engine and the mill. The highest (possibly inaccurate) estimate was 12 bags an hour; and the lowest half a bag. The average was 4·7 bags per hour. The large imported

TABLE 9.8 Investment cost per unit of output (E.A.*s*)
*(equals total initial investment cost divided by hourly output)*

| | *Average* | *Highest* | *Lowest* | *Standard deviation* | *Average expressed as a ratio of hand mill* |
|---|---|---|---|---|---|
| 1. Hand | 4200 | 7500 | 1320 | 3270 (77·9%) | 1 |
| 2. Water[d] | 14,760 | 30,000 | 1070 | 15,400 (104·8%) | 3·5 |
| 3a. Local Hammer | 3620 | 6200 | 2110 | 1580 (43·6%) | 0·9 |
| 3b. Small imported | 4900[e] 5810[f] | 14,860 | 1000 | 3110[e] (63·5%) | 1·2[e] 1·4[f] |
| 4. Roller | 43,130 | 80,000 | 14,000 | 26,700 (61·9%) | 10·3 |

NOTE See Table 9.7 for symbols.

hammer mills showed an average of eight. Output per hour for the roller mills varied from three to 62½; the average was 19½ bags an hour.

Table 9.8 shows the ratio of initial investment cost to hourly output. The absolute figures in Table 9.8 mean very little since they are the total initial investment cost as a ratio to one hour's output. What is interesting is the differences between techniques shown in the last column.

The initial investment cost of a roller mill, in relation to hourly output, is thus ten times that of a hand mill and eleven times that of the local hammer mill. The water mill ratio is higher than that of the hammer mills, and the hand mill. This suggests the water[17] mill may be an inferior technique, since it requires more, not less, investment per man, than these techniques, and has a lower level of labour productivity than the hammer mills – see Tables 9.7 and 9.9.

TABLE 9.9 Output per head (average bags per hour per person employed)

| | | *Standard deviation* |
|---|---|---|
| Hand | 0·07 | 0·05 |
| Water | 0·2 | 0·25 |
| Local hammer | 2·5 | 1·5 |
| Imported | 2·2 | 1·5 |
| Roller | 1·7 | ·05 |

Table 9.9 shows how output per employee varies according to the technique. Taking Tables 9.8 and 9.9 together, it appears that the hand mill maximises employment in relation to investment (and also in relation to output, with the lowest labour productivity), while the roller mill generates least employment in relation to any given investment, and has lower investment *and* labour productivity than the hammer mills. But one cannot draw any conclusions without taking a number of other factors into account: the comparison ignores differences in length of life of the assets, in repair and maintenance, and in requirements for other scarce inputs (e.g. fuel); it assumes equal capacity utilisation between techniques, neglects differences in the type of output, in savings and in income distribution which may vary according to the technique. The rest of this chapter considers these factors.

*Length of life*

It is always difficult to get accurate length of life estimates: lives of assets vary with care, maintenance and repair expenditure, obsolescence due to economic and technical changes, as well as with sheer chance. The evidence suggested that the hand mill (being the simplest machine and suffering least wear and tear) would have the longest life and the hammer mills the shortest, with the roller mill somewhere in between.

TABLE 9.10 Annual investment cost, $\bar{I}$ (average) (E.A.*s*)

| | *Discount rate* (%) 0 | 10 | *Assumed length of life (years)* |
|---|---|---|---|
| 1. Hand mill | 7·0 | 22·3 | 30 |
| 2. Water mill | | | |
| equipment | 42 | 47 | 15 |
| buildings[c] | 35 | 175 | 50 |
| Total[c] | 77 | 222 | |
| 3a. Local hammer | | | |
| equipment | 667 | 1315 | 15 |
| buildings | 210 | 668 | 30 |
| Total | 877 | 1983 | |
| 3b. Imported hammer (small) | | | |
| equipment | 720 | 1420 | 15 |
| buildings[e] | 312 | 1574 | 50 |
| Total[e] | 1032 | 2994 | |
| 4. Roller: | | | |
| equipment | 29,725 | 69,854 | 20 |
| buildings | 4872 | 24,579 | 50 |
| Total | 34,597 | 94,433 | |

NOTES

The figures for initial investment cost are derived from the average acquisition cost shown in Table 9.6.

No evidence was supplied for building lives: the local hammer mill buildings' are assumed to be lower than the rest because of their much lower cost. Water mills have very long lives (three were already over 25 years old, one over 40). But the stones need periodic replacement. Length of lives for the hammer and roller mills (machinery) is the average of the estimates supplied.

See Table 9.7 for symbols.

These estimates are for physical lives and do not take into account economic obsolescence. Length of life estimates make more difference to investment costs (discounted) at low interest rates, than at high. Table 9.10 shows the annual investment cost calculated using two discount rates, 0 per cent and 10 per cent, on the basis of the lives shown in the table. This annual investment cost, described as $\bar{I}$, is calculated by finding the equivalent constant annual payment paid over the life of the asset whose present value is equal to the initial costs of the asset.[18] The comparative costs of the two relatively long-lived techniques, the hand mills and the water mills, are unaffected. The costs of the other three techniques are increased, comparatively; the increase is greatest for the local hammer mill, because of the shorter lives assumed for the buildings. The higher the discount rate the smaller the relative cost increase. Thus with the 0 per cent discount rate the ratio of the investment costs of the local hammer mill to those of the hand mill increases by 60 per cent: with a discount rate of 10 per cent the increase is only 14 per cent. The introduction of length of lives and discounting does not, however, make a substantial difference to the results. The ranking between techniques remains unaltered.

*Repair costs*

These varied greatly between the different examples of each technique. However, repair expenditure on hand[19] and water mills was very low. Repair expenditure on the hammer mills was much more significant. The average cost was 1000/- a year, being a little higher for machines that had been purchased second-hand than those bought new (1235/- compared with 1161/-). There was a definite tendency for repair costs to increase with age, as Table 9.11 shows.

TABLE 9.11 Repair costs of hammer mills

| *Age of machine* | *Average annual repair cost* (E.A.*s*) | *No. of observations* |
|---|---|---|
| Under 5 years | 597 | 23 |
| 5–9 | 672 | 11 |
| 10–14 | 2007 | 12 |
| 15 and over | 2614 | 7 |

In the calculations below, hammer mills are assumed to require 1000/- repairs per year, which, at a 10 per cent discount rate is roughly equivalent to an expenditure of 630/- for the first ten years of life and 2000/- for the following five years. There were substantial variations in estimates for repair costs for the roller mills, ranging from 3500/- p.a. to 600,000/-, but this last included substantial replacement. Allowing for this, 23,000/- p.a. is assumed, as being the rough average.

TABLE 9.12

| | *Assumed annual repairs* (*R*) (E.A.*s*) | $\frac{\bar{I}+R}{\bar{I}}$ | $\frac{\bar{I}+R}{L}$ *as % of roller* | $\frac{\bar{I}+R}{O}$ *as % of roller* |
|---|---|---|---|---|
| 1. Hand | 2 | 109·0 | 0·4 | 8·0 |
| 2. Water[c] | 10 | 104·5 | 3·3 | 24·0 |
| 3a. Local hammer | 1000 | 150·4 | 26·6 | 11·0 |
| 3b. Imported hammer | 1000 | 133·4[e] | 21·9[e] | 12·3[e] |
| | | 120·6[f] | 29·0[f] | 12·5[f] |
| 4. Roller | 23,000 | 124·4 | 100 | 100 |

The repair cost estimates have the biggest proportionate effect on the annual investment cost of the local hammer mill, increasing it by more than 50 per cent, as can be seen in Table 9.12. The low repair costs of the hand and water mills slightly improve their relative performance. However, again the changes are too small to alter the ranking or the general picture.

### CAPACITY UTILISATION

The survey revealed systematic differences in capacity utilisation. The roller mills were far more fully used than the other techniques. The two largest roller mills were operating a 24-hour shift system; one other was working 12 hours a day and intended to extend it to 24 hours. None of the roller mills were working less than eight hours. In contrast the hammer mills were for the most part working two or three hours a day, with a longer day of six to nine hours on market days (generally twice a week). The water mills also were being operated at less than full capacity, though some did work through the night. None of the hammer mills operated a night shift. The hand mills were also out of use for a large part of the time. The differences in capacity utilisation were due to a number of factors. The most important was demand for the products of the mills. Almost all the hammer mills said that lack of demand for their services was responsible for their short hours. That this was so was reflected in the longer hours worked on market days. In contrast the expanding demand for roller mill products justified round-the-clock operations.

Demand by itself cannot explain differences in capacity utilisation, since this is determined by the *balance* of capacity and demand. Thus the hammer mills' capacity was far greater in relation to demand than the roller mills'. There are four factors contributing to the difference in balance: first, it seemed that the industry was not in an equilibrium situation. Roller mill capacity had not caught up with the increases in demand for its products – hence the possibility of profitable 24-hour operation. In contrast, over-capacity in hammer mills in part reflected a lag between changing patterns of demand, away from hammer mill products, and investors' appreciation of and reaction to this change.[20] When investment decisions and available capacity have adapted to the changed demand it seems likely that capacity utilisation of hammer mills will be greater and that of roller mills less. Some of the owners of hammer and water mills noted that their output had declined as consumers switched to roller products. This was particularly marked in mills in and near Nairobi.

Secondly, the small hammer mills provided one of the few openings for small African rural entrepreneurs in Kenya. The total capital outlay was not great, and only relatively simple skills were required. Both were within the capacity of many cash farmers. In contrast, the roller mills required major investment of capital management and skills and was therefore in a completely different class as far as entrepreneurial opportunity is concerned. All the roller mills were Asian and/or European owned and run, whereas the vast majority of hammer mills were in African hands. The sort of minimum return required differed correspondingly, and hence investment, in relation to demand, was far greater for the hammer mills (where the required minimum return was lower) than for the roller mills. Lags were also partly responsible here. It appeared that hammer mills had only been regarded as an African investment opportunity for a few years: this was indicated by the recent date of many of the African investments, and by the fact that older machines, now African-owned, had been bought from Asians in a number of cases. The opportunity to invest in hammer mills had thus appeared simultaneously to a large number of potential African entrepreneurs. They took the ruling returns as indicative of the possibilities: as a result of simultaneous investment, overcapacity emerged and the returns were cut. Many of the hammer mills' owners said that capacity utilisation had been higher before so many mills were installed.

Thirdly, the nature of the market was such that 24-hour operation was easier to achieve with roller than hammer mills. The hammer mills rely for supplies on local farmers (usually within a maximum of two to three miles): these supplies come in baskets and the farmers wait for the maize to be ground and then take it home. The roller mills get all their supplies from the Maize and Produce Board in 200 lb bags. For this reason 24-hour working would be very difficult for hammer mills –

though 10-hour operation would be a possibility. It is also in general easier to organise three-shift working in a few large-scale outfits than in a number of very small ones. Fourthly, the incentive for full utilisation of capacity was greater for the roller mills than for the hammer mills, since the ratio of overhead costs to variable costs was greater.

Some of the factors responsible for differences in capacity utilisation were not essential characteristics of the different techniques but aspects of the particular investment and demand situation in which the industry was placed.

Differences in capacity utilisation make a major difference to the results. Previous tables showed the relative performance assuming equal (8-hour per day) capacity utilisation. The implications of changing this assumption depend on the degree to which employment and output rates vary directly with the hours in operation. It is fair to assume that output varies with hours of operation. Employment varies less than proportionately because of the existence of overhead labour (for management, repairs and maintenance) which will not vary with the time of operation, and also because labour tends to be hired in eight-hour lumps. Table 9.13 shows how the comparison between techniques changes if we assume that water and hammer mills are operated half as long as the roller mill (the hand mill one-sixth as much), and employment and output both vary directly with hours of use.

TABLE 9.13 Capacity utilisation

| | *Assumed utilisation in relation to roller** | $\frac{I}{L}$ *in relation to roller = 100* | $\frac{I}{O}$ *in relation to roller = 100* |
|---|---|---|---|
| 1. Hand | $\frac{1}{6}$ | 2·8 | 57·5 |
| 2. Water | $\frac{1}{2}$ | 7·9[c] | 57·1[d] |
| 3. Hammer (local) | $\frac{1}{2}$ | 43·9 | 18·1 |
| 4. Hammer (imported) | $\frac{1}{2}$ | 40·8[e] | 22·9[e] |
| | | 59·8[f] | 25·9[f] |
| 5. Roller | 1 | 100·0 | 100·0 |

* assumed to operate 16 hours per day.

The changed assumption with respect to capacity utilisation much reduces the relative advantage of the hammer (and water and hand) mills in comparison with the roller mill. If we assumed that the roller mill operated for four times the hours per 24 hours as compared with the hammer mills, the investment cost per worker on the local hammer mills would be very nearly as high as the roller mill, on the assumption of a proportionate change in employment – though the investment costs per unit of output would still be substantially smaller (36 per cent). But the assumption of proportionate change in employment is illegitimate because a fairly large proportion of employment on the roller mills does appear to be overhead – say between $\frac{1}{3}$ and $\frac{1}{2}$; the hammer mills provide employment by the day, and hence this employment does not vary if actual use of the machine is less than a full day; and the figures for employment provided by the hammer mills already reflected, to some extent, their low capacity utilisation as many of them quoted higher figures (sometimes double or more) for 'busy' periods. For these reasons the comparison given in Table 9.13 would seem to give a fair guide to the implications of differences in capacity utilisation since the possible overestimation of relative utilisation of the hammer mills is offset by overestimation of the variability of employment with respect to capacity utilisation. Investment costs per unit of output are more likely to vary directly with capacity utilisation: with four times the capacity utilisation of the roller mill, as compared with the hammer mill, investment costs per unit of output of the hammer mill would be between one-third and one-half of those of the roller mill. It is difficult to know what assumptions one should make about capacity utilisation in choice of technique. To the extent that the difference in capacity utilisation reflects a non-equilibrium situation, as argued above, it should not be assumed in more general analysis, particularly if the conclusions are to be used for other periods and/or other countries. On the other hand, for short-term policy purposes, the existence of considerable spare capacity among the hammer mills but not the roller mills suggests that investment could be saved simply by shifting additional demand towards the products of the hammer mills and thus avoiding the need for new investment.

*Use of other scarce resources*

The two main resources used in addition to investible resources and unskilled labour were skilled labour and fuel (including water).

*Skills:* skilled labour may be required for operation, maintenance or repair. Table 9.14 shows the typical skill requirements for these operations.

Skill requirements increased with the investment-intensity of the technique. Only the roller mills require skilled labour for operation of the mill. The level of skills, as well as the number of skilled workers employed, was also higher for the operation of roller mills than for the diesel mills. The repair activities of the diesel mills did need some mechanical skills, but these were almost invariably obtainable locally,

TABLE 9.14 Skill requirements

| | *Operation* | *Maintenance* | *Repair* |
|---|---|---|---|
| Hand | none | none | v. little |
| Water | none (2–3 week training) | v. little | little |
| Diesel | none (2–3 week training typical) | v. little | skilled mechanic, normally from outside firm |
| Roller | substantial proportion of workers: in one case 54% of workers ex. packers | | requires skilled labour, normally from within the firm |

sometimes provided by Asians, sometimes by Africans. In contrast the large roller mills employed a number of Europeans for some of their more general supervising and repair operations. The one large roller mill which provided a detailed breakdown of their labour force employed one engineer with four or five years' experience and training in the U.K. Quantification of skills used is always tricky, partly because of the question of what valuation to place on them. In this case difficulties are made worse by lack of systematic information on skills plus salaries, and because part of the cost of skills used has already been included in the figures for repairs. No attempt is therefore made here to quantify further.

*Fuel:* 'blood' was the graphic description of fuel required for the handmill. The other methods used more prosaic fuels. Fuel estimates varied quite a lot, sometimes because of the assumption made about capacity utilisation. Table 9.15 shows what seems to be a typical fuel cost per bag ground.

TABLE 9.15 Fuel cost per bag (E.A.*s*)

| | |
|---|---|
| Hand | none |
| Water | 45/- p.a.≃20 cents* |
| Diesel | 0/75 |
| Roller | 0/75–1/- |

* assumes about 6 bags ground per week.

Including fuel costs slightly improves the relative position of the hand and water mill, and leaves the comparison between diesel and roller mill more or less unchanged.

*Quality of output*

The total costs per bag of the roller mills are considerably greater than the other mills. How much greater depends on relative capacity utilisation. However, the quality of the product also differs, as discussed earlier, and although the sifted product is nutritionally inferior to the other products it is sufficiently preferred by some consumers to lead them to pay a much higher price for it – 40 cents a lb as compared with 25 cents a lb. This difference in price includes the cost of the raw material. The difference in grinding charge is proportionately greater – the official charge for sifted is 410 per cent of that for ordinary posho. In the view of these consumers, then, the higher costs are justified by the difference in quality. If one weights the sifted output as worth four times as much as the products of the hammer mill, then the investment/output ratio of the roller mill falls in relation to that of the other mills: for example, the 'adjusted' $\bar{I}/O$ of the roller mill becomes 1·4 times that of the local hammer hill, whereas without such an adjustment it is 5·5 as great. Labour productivity of the roller mill would be correspondingly increased. This adjustment is not big enough to alter the basic comparison, but it does modify it, very severely. The question of whether one should make such an adjustment is of wider significance. The extra payment consumers are prepared to make for the sifted flour reflects a particular distribution of income, and is in part a consequence of fairly heavy advertising conducted by the producers of the flour, who in their own words 'created the market'. The flour is nutritionally inferior.

*Savings levels*

Savings may be affected by choice of technique in two different ways: first, through the direct surplus generated as a result of the technique – the Galenson-Leibenstein-type effect; and secondly, because people may be induced to save because of the investment opportunities open to them. The latter effect has been generally ignored, as it is inconsistent with the normal assumption of separation of investment and savings decisions.

Savings directly generated are normally identified with profits, assuming all profits are saved and all wages consumed. Then for a given amount of expenditure on investment, direct savings are equal to the value added by the investment less the amount paid out in wages. The value added (in monetary terms) is regulated by the Kenya Government: prices vary somewhat according to area and quantity. In 1969 the regulated charge for ordinary posho was around 3/25; and for sifted grade I, 14/00. In fact the prices charged by the hammer mills diverged a bit from this price. Their average, for ordinary posho, was 4/-.

*Wage-rates*

The normal assumption made, in the literature of choice of technique, is that the wage-rate is the same irrespective of the technique. However, the survey revealed considerable variations between wage-rates, as shown in Table 9.16.

TABLE 9.16 Wage-rates according to technique

| | *Daily rate* | *Monthly* | *Combined assuming 20-day month* |
|---|---|---|---|
| Water | 1/20 | 62/- | 45/- |
| Hammer | 4/10 | 114/- | 99/- |
| Roller | 9/60 | 354/- | 293/- |

Some firms provided information on daily rates, some on monthly, and a conversion rate of 20 days per month has been used above.

Wage-rates are considerably higher in roller mills than hammer mills, and hammer mills than water mills. Since the hand mills are not generally used commercially, and always on a self-employment basis, they have been excluded. The difference in wages is due to a combination of three factors, working in the same direction: differences in skill requirements (with the roller mill requiring more skilled labour and therefore high-wage labour, as noted above); differences in scale, which in general seem to be associated with wage-rates, as noted in the last chapter; and differences in location. Urban wages for both types of mill were substantially higher than rural wages; and, as noted earlier (Table 9.4), the roller mills were all located in towns while a very large proportion of the hammer mills were in the rural areas. Table 9.17 shows the difference in wages according to location.

TABLE 9.17

| | *Hammer* | | *Roller* | |
|---|---|---|---|---|
| | *Daily* | *Monthly* | *Daily* | *Monthly* |
| Rural areas | 3/90 | 78/50 | — | — |
| Small towns (1000–10,000) | 4/75 | 120/00 | 5/- | n.a. |
| Large towns (over 10,000) | 6/75 | 239/30 | 12/50* | 35/- |

* includes highly skilled.

The marked differences between urban and rural wage rates is partly a reflection of government policy towards minimum wages, and the possibility of enforcing such policies. It may also be in part due to differences in the organisation of labour, and in the balance of supply and demand. In particular rural employees are more likely to be doing some farm work as well; a living wage is not therefore as essential for them as for urban workers.

Table 9.18 estimates the surplus generated by each technique. On the assumption of equal (8-hour) operation by each technique, the hammer mills generate considerably larger surpluses than the alternatives – the local hammer mill being associated with four times the surplus of the roller mill. However, equal capacity utilisation is a key assumption, as shown by the calculations on alternative assumptions. While these alternative calculations are crude, not making allowances for overhead costs other than capital, and assuming that the surplus moves proportionately with capacity utilisation,[21] the general picture is clear. As the relative capacity utilisation of the roller mill increases, its relative surplus also increases: if hammer mills are assumed to operate for four hours a day and the roller mills for 16, the surplus is about the same. If the roller mills operate for 24 hours a day, and the hammer for four, then the surplus of the roller mills exceeds that of the hammer mills. In practice, the most efficient of the roller mills were operating for 24 hours a day, while many of the hammer mills were operating for four hours a day or less. Thus the roller mills in general were probably generating a greater surplus than the hammer mills. This is borne out by the impression that the roller mills offered greater profits in relation to investment than the hammer mills. This, it must be emphasised, is an aspect of the market situation, and the consequent relative capacity utilisation, not the technical characteristics of the techniques.

The surplus, as defined here, does not entirely or automatically go to savings. In the first place it is required to meet overhead expenses, such as administrative costs. The impression received, without detailed figures, is that administrative and other overhead costs were substantially greater for the roller mills than for the other techniques. It cannot, of course, be assumed that any surplus available after deducting overheads is saved. For the self-employed sector (which included most of the hammer mills) the surplus went partly towards the living expenses of the owner. For the roller mills the surplus also takes the form of income for the owners and may be spent, or remitted abroad. In view of these possibilities, and the fact that it is not clear which technique maximises surplus, since it depends on capacity utilisation assumptions, all that can safely be said is that the maximising surplus criterion of choice of technique does not, as is often assumed, automatically point to the most investment-intensive technique. In the case under consideration there is some conflict between employment and growth in that the

TABLE 9.18 Calculation of surplus generated per 100,000/- initial investment

| *Technique* | (1) *Water* | (2) *Hammer (local)* | (3) *Hammer* (small imported)* | (4) *Roller* |
|---|---|---|---|---|
| 1. Hourly output (from Table 9.8)—bags | 6·8 | 27·6 | 20·4 | 2·3 |
| 2. Daily output with 8-hr. day | 54·2 | 221·0 | 163·3 | 18·6 |
| 3. Price charged per bag† (E.A.*s*) | 3/90 | 4/00 | 4/00 | 14/00 |
| 4. Fuel cost per bag (Table 9.15) | 0/20 | 0/75 | 0/75 | 0/80 |
| 5. Approx. repair cost per bag | 0/30 | 0/12 | 0/10 | 0/40‡ |
| 6. % wastage per bag ground | 3·2 | 0·5 | 0·5 | 20·0 |
| 7. Cost of wastage§ | 0/90 | 0/14 | 0/14 | 2/80 |
| 8. Net value added per bag | 2/50 | 2/99 | 3/01 | 10/00 |
| 9. Net value added per day | 135/5 | 672/75 | 491/53 | 186/00 |
| 10. Employment (from Table 9.7) | 51·0 | 11·9 | 12·0 | 2·4 |
| Wage rate (daily) | 2/25 | 4/95 | 4/95 | 14/65 |
| 11. Wage bill (daily) | 114/75 | 58/91 | 59/40 | 35/16 |
| 12. Surplus: 8-hr. shift | 20/75 | 613/84 | 432/13 | 150/84 |
| 13. Alternative (*a*) assumption about shift | 8-hr. | 8-hr. | 8-hr. | 16-hr. |
| 14. Surplus on assumption (*a*) | 20/75 | 613/84 | 432/13 | 301/68 |
| 15. Alternative (*b*) | 4-hr. | 4-hr. | 4-hr. | 24-hr. |
| 16. Surplus on assumption (*b*) | 10/38 | 306/92 | 216/07 | 452/52 |

* Using low estimates for imported hammer mills.

† This is charge for grinding and does not include raw material cost.

‡ If repair costs were the same whether the machines were operated for eight hours or 16 hours then they would work out at 60 cents per bag. Here it is assumed that there is some saving for eight-hour operation, so repair costs are 40 cents per bag.

§ In the case of the roller mill the wastage cost has been halved to allow for some recovery by using the discarded elements as cattle-feed.

NOTE Surplus calculations on alternative assumptions about capacity utilisation have assumed all costs change proportionately with output: consequently the surplus is also changed proportionately. This is unrealistic – some overhead labour and repair costs will not change proportionately. On the other hand the employment figures for the hammer already reflect less than full capacity utilisation.

techniques which maximise employment – the hand mill, which involves no monetary surplus, and the water mill – are associated with less surplus than the hammer mill. The hammer mill appears to have a comparable surplus to the roller mill except under extreme assumptions about capacity utilisation, while being substantially cheaper per person employed. Savings may also be stimulated indirectly: the provision of small-scale investment activities seems to stimulate local savings to take advantage of the opportunities. This was indicated by the large number of hammer mills which had financed their activities out of their own savings. Though obviously this is not conclusive proof it suggests that small-scale investment opportunities may, to some extent, generate their own savings.

*Income distribution*

The techniques generate incomes for quite different groups within society. Broadly, the income from the roller mills goes to industrial workers, and provides them with a much-above-average income, and to the owners and shareholders of the large companies which own the mills. It also allows considerable incomes for salaried employees, both local and foreign. The water and hammer mills provide income for rural workers and very small-scale entrepreneurs, mostly in the rural areas and local mechanics, for repairs. The hand mill provides activity and income (in kind) in the subsistence sector.

*Summary and conclusion*

This chapter has examined choice of technique in maize grinding on the basis of a survey conducted in 1969. Maize grinding presents an interesting example because it is concerned with processing one of the most basic elements of diet in Kenya. The methods used differ in location, scale, power, labour-intensity (defined as $I/L$) and labour productivity. The quality of the product also differs.

The evidence from the survey shows that the most investment-intensive technique, the roller mill, requires nearly 200 times as much investment per worker as the most labour-intensive method, the hand mill; over ten times as much as the water mill; and nearly 5 times as much as the local hammer mill, if we assume equal capacity utilisation for each technique. It also involves more investment in relation to output ($I/O$) with ten times the investment cost per unit of output of the hand mill and twelve times that of the local hammer mill. Adjusting the investment cost to allow for different lives of assets, and for repair costs, does not alter the general conclusion. The labour productivity of the most investment-intensive technique is lower than that of the hammer mill, but higher than that of the hand mill and water mill, if labour productivity is defined in terms of bags ground. It would thus appear that the roller mill is technically inefficient or inferior as compared with the hammer mills, requiring more investment *and* more labour in relation to output; and indeed this also appears from the price charged per bag ground, which is 3–4/- for the hammer mills, and 14–15/- for the roller mills. The roller mills require imported machinery, whereas the mill part of the hammer mills is now locally produced. The roller mill is essentially an urban large-scale technique of production, whereas the hammer mills are generally rural and small-scale and pay lower wages. From the point of view of equality and of rural/urban balance the hammer mills are to be preferred. Despite all this, demand for the products of the roller mills is rapidly rising, while that for the hammer mills is falling. This trend arises from the fact that the consumers who possess the bulk of the monetary purchasing power prefer the whiter flour produced by the roller mills; the preference is partially (but by no means wholly) a reaction to advertising, which is concentrated exclusively on the products of the roller-mill. The switch also arises from the urbanisation process, since the roller mill products are more suited to an urban society, being better packed and keeping longer. Hammer mills are most suited to providing fresh flour for those who grow and consume their own flour, i.e. rural consumers. The switch in consumption patterns is also partly due to the rise in urban incomes – itself a reflection of the kind of choice of technique of which the roller mill is an example – which involves relatively high-wage urban employment. The trend towards consumption of the products of roller mills, together with time-lags, and a shortage of investment opportunities for small-scale entrepreneurs, has meant that capacity utilisation of the roller mills is substantially greater than that of the hammer-mills. The difference in capacity utilisation significantly modifies conclusions about the investment costs of different techniques, and also their relative employment-generating implications. Assuming that hammer mills operate half the time that roller mills do, their investment costs per unit of output are only 18 per cent of those of the roller mill, while their investment costs per employee are 70 per cent. More extreme differences in capacity utilisation would eliminate the cost advantages of the hammer mills. Capacity utilisation assumptions were also of critical significance in determining the surplus generated. With equal capacity utilisation the hammer mills generated over four times as much surplus for a given amount of investment. But if the capacity utilisation of the hammer mill falls much below one-quarter of that of the roller mill then the roller mill maximises the surplus.

The conclusion has been largely phrased in terms of the choice between hammer and roller mills. A distinction was made throughout the tables between the local and imported hammer mill. This distinction simply reflected differences in the costs and standards of buildings which housed the two. Which figures apply to future investment decisions depends on the type of building adopted. Clearly the local mill – which had identical performance with the imported mill but used less expensive buildings – is to be preferred in any choice, assuming that the cheap buildings continue to be associated with it. The water mill in many cases does not offer a genuine alternative because it requires the right sort of water flow for operation. Unless it is very fully used, its investment productivity is lower than that of the hammer mills. Where the right sort of water is available at low cost it is an alternative, but not elsewhere. The hand mill serves a different set of consumers – the subsistence sector: its investment cost in relation to output (assuming it is only operated one-third as long as the hammer mills) is about twice that of the hammer mill; and it is also highly fatigue-intensive.

If one weights each bag ground as equivalent to every other bag ground – i.e. gives no additional value to the sifted product – then the hammer mills are to be preferred as a technique on grounds of investment cost, employment generation, rural/urban balance, income distribution, nutritional consequences, and use of local machinery. This remains true, assuming that the hammer mills operate at half the capacity of the roller mills. The social costs imposed by the current trend away from ordinary to sifted flour are greater than the straightforward comparison of costs discussed would suggest. This is because full capacity in the roller mills and low capacity utilisation in the hammer mills mean that a further switch from hammer to roller mills would involve new investment in roller mills, while it would not save investment in the hammer mills but further reduce capacity utilisation. Equally, a reverse switch would involve no new investment but simply allow greater capacity utilisation in the hammer mills, in the short run.

Despite the apparent superiority of the hammer mill, roller mills were clearly the most profitable technique: this arose because of 'consumer preference' for their product. Taking this 'preference' into account and weighting the value of output according to its monetary value severely modified the conclusions: the investment productivity of the hammer mill was then only slightly higher than that of the roller mill, while the labour productivity of the hammer mill fell below that of the roller mill.

The ultimate conclusion, then, depends critically on the value placed on consumers' sovereignty. Most of the assumptions needed to justify consumers' sovereignty are questionable in this case. Advertising is heavily concentrated on one of the products only, whose producers claim explicitly to have 'created the market'; the assumption that consumers are the best judges of their own interests may be challenged given the very substantial nutritional inferiority of the sifted product;[22] finally, the pattern of demand is heavily influenced by an unequal income distribution. The 'consumer preferences' exhibited in the market are the preferences of those with purchasing power – hence the use of inverted commas above.

NOTES

1. It is the staple food in all the maize-growing areas, but not on the Coast, or for non-cultivating tribes such as the Masai. One of the witnesses to the Maize Commission of Inquiry estimated that sisal workers consume three bags per year, or 1½ lb per day, while each dependant consumes half as much. (Evidence of R. H. Daubeney, *Maize Commission of Inquiry*, Vol. 1, 2nd Day, 11 January 1966.) Evidence for some villages in North-east Tanzania illustrates the importance of maize in the diet of villagers. Studies of three villages showed that maize provided 30–50 per cent of total calories, 20–40 per cent of protein, 10–30 per cent of iron consumed. See W. Poeplau and C. Schlage (1966).
2. Estimates derived from evidence to Maize Commission of Inquiry, January 1966. G. K. Kariithi, Permanent Secretary to the Ministry of Agriculture, estimated that total production was normally 13m. bags of which 1·6m. were marketed. J. Block, Deputy President of the Kenya National Farmers' Union, estimated production at 16½–17m. bags, of which the maize board handled 2m. Blundell (7th Day, 18 January 1966) estimated production at 13½–15m. bags, of which nearly 1m. were produced on large farms. Since 1966, partly as a result of the success of hybrid maize, the quantity coming on the market has increased substantially. It was 3.6m. bags for 1967–8. It does not necessarily follow that total production has increased proportionately – with the gradual extension of the cash economy and specialisation one would expect total production to increase less than the marketed crop.
3. From Kariithi's evidence to the Maize Commission of Inquiry.
4. Lists of millers were supplied by the Maize and Produce Board, District Commissioners and County Council officials. The very large number of mills in Kenya was indicated by the size of the lists submitted. For example, the Bungoma County Council supplied a list of 75 millers for Bungoma district: the D.C. of Nandi District a list of 88. The survey was conducted by interview carried out by myself and two assistants, students at the University of East Africa.
5. Schlage (1968), p. 6–8.
6. One of the witnesses to the Maize Commission of Inquiry gave a similar description of traditional methods: see Chandaria's evidence, Vol. 4, 12th Day, p. 27.
7. Though they do, of course, involve some investment in the sense that there is some delay between the labour input and the final product. While this should undoubtedly qualify as investment, where delay or roundaboutness are taken to be the essential characteristics of investment, it is ignored here. In the first place, this type of investment is akin to working rather than fixed capital, and working capital has been excluded from this survey. In the second place, we are concerned to assess investment-intensity because investment is assumed to represent a scarce resource, in contrast with (unskilled) labour, which is assumed abundant. Investment is a scarce resource in so far as it makes use of scarce foreign exchange, or uses part of the country's limited capacity to produce capital goods. The 'capital' involved in the subsistence sector is a matter of using labour at one time, say $t - 1$, for the production of a product at a later period, $t$. If labour is abundant at $t - 1$ and $t$, then this type of investment does not represent the use of a scarce resource. If the workers employed at $t - 1$ have to be paid at that time, there may be some opportunity cost as a result of the additional consumption at $t - 1$. This is why wages advanced to workers are normally included as part of working capital. But in the subsistence sector no such advance is paid.
8. Evidence to Maize Commission of Inquiry, Vol. 1, p. 47 (my italics).
9. Maize Commission of Inquiry, Vol. 6, p. 22.
10. The Hyatt Report, 1968.
11. Maize Commission of Inquiry, Vol. 4, p. 27.
12. Ibid., Vol. 4, p. 41.

13. Ibid., Vol. 5, p. 40.
14. Ibid., Vol. 4, p. 38.
15. Vol. 4, p. 34.
16. E.g. Boon (1964) uses 'average' figures without indicating variations between examples of the 'same' technique; Sen (1968) similarly provides a single set of characteristics for each technique in his analysis of cotton weaving.
17. In Table 9.8 the figures include the expensive dam, but as this mill had exceptionally high output levels its inclusion makes very little difference to this ratio.
18. The formula for converting a constant annual cash flow, $A$, into present value is well known:

$$PV = \frac{A}{r}\left[1 - \frac{1}{(1+r)^n}\right]$$

where $r$ = interest rate and $n$ = number of years.

In this case the inverse operation is performed, and an initial acquisition cost, which may be thought of as the present value of the capital cost, is converted into a constant annual cash flow, $A$, by dividing the initial cost by

$$\frac{\left[1 - \frac{1}{(1+r)^n}\right]}{r}$$

19. Of the five owners of hand mills two had spent nothing on repairs, one of which was for a mill 18 years old; one had spent 11/- in 14 years, another 20/- in 11 years, and one estimated an annual expenditure of 10/-.
20. With downward shifts in demand there is normally a tendency for a greater lag before equilibrium capacity utilisation occurs, than with upward shifts, because of the asymmetry between investment and disinvestment. Since bygones are bygones, it may be worth keeping plants open, despite industry excess capacity, so long as variable costs are covered. Since upward shifts may be adjusted to as soon as they are recognised, apart from a gestation lag, the asymmetry is itself enough to explain economy-wide spare capacity in any economy in which demand patterns shift continuously and unpredictably, even in the absence of all the other factors that contribute to the existence of spare capacity.
21. Because some of these costs are overhead the surplus can be expected to increase more than proportionately with increases in capacity utilisation (and decrease more than proportionately with decreases), reinforcing the above conclusions.
22. The nutritional inferiority arises partly from the loss of nutrients, discussed earlier – see Table 9.2. These *can* at considerable expense be 'put back', a similar operation to that which is carried out on white bread, for example, in advanced countries. But in addition the reduced bran in the diet is thought to present serious health hazards: to put this back is equivalent to going back to the non-sifted product.

# 10 Cement Block Manufacture in Kenya

Cement blocks are, probably, the most important building material in urban Kenya. They are made by mixing cement, sand and small stones together and forming the mixture into blocks of varying size. The blocks are then used as bricks in the construction of buildings. Block manufacture was chosen because it was known that various different ways of making blocks were in use in Kenya. This study primarily focuses on these different methods. But cement blocks are just one building material among many. Some of the alternative building materials and the implications for choice of technique are considered at the end of the chapter.

## THE SURVEY

Twenty-three different organisations were interviewed, each operating one or more block-making machines. They were not intended to be a representative sample of block-makers in Kenya. The aim was to find a number of examples of each of the different ways of manufacturing blocks, rather than to give any sort of accurate representation of the industry in Kenya as a whole. Of the 23, 19 were located in or very close to Nairobi. The remaining four came from Thika, Machakos, Nyeri and Nanyuki. The 19 in Nairobi included all the main block suppliers, but the rest of the country was extremely sketchily represented. The organisations to be interviewed were selected in a somewhat *ad hoc* way from among the main types of organisation known to use block-making machines – viz. quarry operators and other building material firms, builders and local government authorities.[1]

The 23 covered a number of different types of organisation: ten operated quarries and thus provided an outlet for some of the products of the quarry while benefiting from the proximity of the most bulky raw material for block-making – ballast or small stones. Eight were building firms. Whereas the quarry-operator/block-makers produced blocks for sale, all the builders produced exclusively for their own building operations. Most of them relied on commercially sold blocks for a good deal of their operations – primarily those in Nairobi – using their block-making machines only for out-of-town jobs. Thus their requirements, in terms of scale of output and mobility of the machine, differed substantially from those of the main commercial producers, which had important implications for choice of techniques. Three firms were suppliers of building materials (but not quarry-operators) and also made blocks for sale. One of these was a major cement producer, which provided an obvious link with block-making. Another produced building materials such as tiles and had, possibly temporarily, stopped production of cement blocks because it was less profitable than its other products. Representatives of two councils were interviewed: Nairobi City Council and Machakos County Council. Like the builders, they produced for their own use exclusively. In Nairobi all the blocks for a major housing scheme were being produced, and the Machakos block-makers provided virtually all the blocks used by Machakos County Council. Seventeen of the organisations were Asian-owned and run: the remainder, with the obvious exception of the two Councils, were European. All the quarries were Asian. The interviews were carried out at the time when the Kenyanisation policy and the Asian exodus were in full flood. This had some relevance to the questions under survey since it appeared that in some cases short-lived assets were being chosen in preference to long, mainly because of the peculiar uncertainties at this time.

The 23 organisations between them owned and operated forty block-making machines. These machines could be classified into three broad categories:

Hand-operated machines;
Electrically or diesel-powered vibrating stationary machines;
Electrically powered mobile laying machines.

Within each of these categories (which will be described in more detail later) there were significant distinctions. The hand-operated machines were basically of similar design, though they differed in age, in cost when bought, and in manufacturer. Within the second category there were major differences in characteristics between imported machines and locally manufactured machines in terms of price, length of life and repair and maintenance requirements. This category has therefore been split into two, locally manufactured and imported machines. Within the latter category there were substantial differences between the machines according to the scale of production for which they were designed. Where scale is relevant the category has been split accordingly. The housing scheme in Nairobi used a locally produced machine which had been especially adapted to the needs of the scheme according to the design and under the supervision of the Resident Engineer. This machine therefore is in a category of its own. Each of the machines in the third category (the laying machines) differed in age and scale of production. Thus generalisation about the group as a whole is not very meaningful.

In the rest of this chapter the techniques have been categorised as follows:

| *Category* | *Technique* |
|---|---|
| 1 | Hand-operated |
| 2 | Locally produced stationary vibrating, of which: |
| 2a | normal machine |
| 2b | special adaptation |
| 3 | Imported stationary vibrating, of which: |
| 3a | small |
| 3b | large |
| 4 | Laying vibrating, of which: |
| 4a | small |
| 4b | large |

TABLE 10.1 Users, location and origin of the machines covered[1]

| *Item* | *Categories* 1 | 2a | 2b | 3[2] | 4 | All |
|---|---|---|---|---|---|---|
| Type of organisation: | | | | | | |
| Quarrier | 4 (2) | 8 (5) | — — | 7 (6) | 2 (1) | 21 (14) |
| Builder | 7 (6) | 3 (3) | — — | 2 (2) | — — | 12 (11) |
| Council | 2 (1) | — — | 1 (1) | — — | — — | 3 (2) |
| Other | — — | — — | — — | 1 (1) | 3 (2) | 4 (3) |
| Location: | | | | | | |
| Nairobi | 7 (6) | 10 (7) | 1 (1) | 9 (8) | 3 (2) | 30 (24) |
| Other | 6 (3) | 1 (1) | — — | 1 (1) | 2 (1) | 10 (6) |
| Origin of machine: | | | | | | |
| Imported | 12 (8) | — — | — — | 10 (9) | 5 (3) | 27 (20) |
| Local | 1 (1) | 11 (8) | 1 (1) | — — | — — | 13 (10) |
| *Sample total* | *13 (9)* | *11 (8)* | *1 (1)* | *10 (9)* | *5 (3)* | *40 (30)* |

[1] Some of the organisations operated more than one machine. The figures in parentheses show the number of organisations. Since some organisations were responsible for more than one type of machine, the total of the figures in parentheses is greater than the total number of organisations.

[2] Category 3 includes both large and small imported vibrating machines. All the large machines (of which there were five) were located in Nairobi; quarries operated four of them, a builder the fifth.

Table 10.1 indicates that quarries account for fewer hand-operated machines (30·8) per cent than for all types of machines (52·5 per cent). They are particularly well represented in categories 2 and 3, the vibrating stationary machines. It also suggests that firms in Nairobi generally use proportionately fewer hand-operated machines than firms outside Nairobi; 54 per cent of hand-operated machines were in Nairobi compared with 85 per cent of the electrically or diesel-powered machines. If we include the one Mombasa firm interviewed with the Nairobi firms as 'urban' firms, this conclusion is reinforced, with over 93 per cent of the powered machines being urban as compared with 54 per cent of hand-operated machines. Table 10.2 gives a breakdown by type of organisation separately for Nairobi and outside Nairobi:

TABLE 10.2 Location of the techniques sampled and kind of organisations using them

| *Location and kind of organisation* | *Machine categories* | | | |
|---|---|---|---|---|
| | *1* | *2 and 3* | *4* | *All* |
| In Nairobi: | | | | |
| Quarriers | 0 | 14 | 2 | 16 |
| Builders | 7 | 5 | — | 12 |
| Other | 0 | 1 | 1 | 2 |
| All | 7 | 20 | 3 | 30 |
| Outside Nairobi: | | | | |
| Quarriers | 4 | 1 | — | 5 |
| Other | 2 | 1 | 2 | 5 |
| All | 6 | 2 | 2 | 10 |

[1] No builders were interviewed outside Nairobi.

No quarries used hand-operated machines in Nairobi, whereas each of the quarriers outside Nairobi did. In Nairobi builders were the only organisations to use hand-machines. All the builders operating in Nairobi used hand-operated machines only for small or out-of-town jobs. As we shall see, this major difference between urban and rural choice of technique, which is paralleled elsewhere[2] is to be attributed to questions of scale, product requirements and relative factor costs.

Although, altogether, the organisations contacted were directly responsible for 40 machines, data have not been included for all 40 in what follows. In some cases firms owned identical machines and gave identical answers to all questions for two or more machines. In such cases the answers have been treated as applying to a single case. In many interviews incomplete answers were obtained – e.g. no estimate of repair costs, or of mixture used. In general, incomplete answers have been included except where virtually no information, beyond a statement of the existence of the machine, was provided. The detailed data analysed in this paper cover 33 cases, broken down as follows:

| *Category* | *No. of examples* |
|---|---|
| 1 | 10 |
| 2a | 8 |
| 2b | 1 |
| 3a | 4 |
| 3b | 5 |
| 4 | 5 |

BLOCK-MAKING MACHINES

Block-making is a process of mixing materials – cement, sand and ballast – and then forming the mixture into blocks. The block-making machine, the subject of this investigation, is the machine which converts the mixture into blocks. Hand-operated machines consist basically of a box the size of the block; the mixture is put into this block, a lid is lowered on to the mixture which is compressed by the pressure of the lid (sometimes lowered several times), which normally has some system of springs to increase the pressure applied. In some examples of hand-operated machines the mixture was hit with sticks to compress it before putting the lid on. After the mixture has been compressed the block is removed and left to dry for two weeks or so, after which it is ready for use. Blocks can be of various sizes and may be hollow or solid. Generally the hand-operated machines were less versatile than the mechanical ones, though some of them could do more than one size, and hollow or solid blocks. Hollow blocks have two advantages over solid ones: they use less material and are lighter to carry. But they are also inclined to be weaker, and are definitely more difficult to make and collapse more often.

With the stationary vibrating machine the mixture is put into moulds (which can be changed to form solid or hollow blocks of different sizes): the machine then vibrates vigorously for about 30 seconds; the blocks are then removed and left to dry. The smallest vibrating machines take one block of 9 × 9 × 18 inches at a time or two of 6 × 9 × 18 inches (described respectively as 9 × 9 or 6 × 9 in what follows). The rate of output thus varies substantially according to the size of the block. The hand-operated machines, mostly, can only take one block at a time.

Hollow blocks take longer than solid because of the higher rate of breakage and the greater care required.[3] The locally produced stationary machines were all designed to produce one 9 × 9 block or two 6 × 9 at a time. Within the imported category some were also of this size while others produced at roughly twice the rate with two 9 × 9 or four 6 × 9 at a time. Where this difference affects the results category 3 has been split into 3a and 3b, small and large imported stationary vibrating machines, respectively.

Laying machines also operate on the vibrating principle, but they are mobile and as they move they lay rows of blocks on the ground, as chickens lay eggs (though not in rows!). The number they lay in each row depends on the size of the machine – which varies substantially – and the size of the blocks. Daily output also depends on the speed of operations or the number of rows laid per day. Since they lay the blocks directly on the ground it is not necessary to transport the blocks from machines to drying area as with the stationary machines. But the mobility also creates problems. The mixture has to be transported to the moving layer; and the ground on which the blocks are laid has to be very smooth if a smooth block is to be achieved. But pallets, on which the blocks from stationary machines are formed and carried away, are not needed. Pallets cost between 2/- and 4/50 each and 2000–4000 are needed.

The process of vibrating not only saves the use of labour to exert the necessary pressure, but also produces a stronger block of more uniform quality. Blocks produced by hand-operated machines, it was claimed, are of more erratic quality: some easily met the minimum requirements of 400 lb crushing strength a square inch, whereas others did not. The tendency for weaker blocks[4] can be compensated for by increasing the proportion of cement used in the block. As this is by far the most expensive material[5] this increases the cost of the blocks substantially. An official in the materials department of the Ministry of Public Works suggested that the correct proportions for cement blocks is one part cement to nine of other ingredients (three of sand and six of stones), but that ratios of 1:12 (which produce a weaker block) are quite common.

Table 10.3 shows the ratio of the mixture adopted by those firms which supplied this information. There is no obvious tendency for a stronger mixture to be adopted when hand machines are used, though one firm which had operated both types of machines used 1:12 for hand machines, 1:15 for vibrating machines. The required strength of a block depends on how much weight it is going to have to support. For one-storey buildings the required strength is less than for multistorey buildings. All the hand-operated machines were producing for one-storey buildings only: most argued that hand-operated machines were not suitable for multistorey buildings, though one operator claimed that they would be suitable if the mixture were strengthened. The mixture

TABLE 10.3 Ratio of cement to other materials in cement blocks in individual cases, by category of block-making machine

| *Category of machines* | | | | |
|---|---|---|---|---|
| 1 | 2a | 2b | 3 | 4 |
| 1:7[1] | 1:6[2] | 1:11 | 1:5[3] | 1:8 |
| | 1:10[4] | | | |
| 1:9 | 1:7 | — | 1:7 | 1:8 |
| 1:10[5] | 1:8 | — | 1:10 | 1:8 |
| 1:12 | 1:8 | — | 1:11 | 1:12 |
| 1:12[6] | 1:12 | — | 1:12 | 1:12 |

[1] This was a double-storey building and included stones. For a single-storey building the firm in question used a ratio of one part of cement to six parts of dust plus sand, which it obtained free of charge. Two hand machines were used in this case.

[2] No ballast.

[3] This very strong mixture was justified explicitly. It was claimed that it produced a very strong block, and that breakages were only 2 per cent, compared with an estimated 15 per cent for a lighter mix. Since the firm concerned was a builder producing blocks for own use, it may have been more concerned about avoiding damaged blocks than other firms.

[4] With ballast.

[5] Apart from cement, the mixture is described as consisting entirely of sand or whatever local materials are available.

[6] A local electric vibrating machine had also been used, with a mix weakened to 1:15.

used by the modified local machine on the Nairobi housing scheme was particularly interesting in this connection. One of the modifications the engineer introduced was to have *double* vibrating on the grounds that a weaker mixture could then be used to secure the same strength block. He had carried out a number of experiments showing how the uniformity and predictability of block strength increased with the increasing rate of vibration. He also claimed that the modifications secured a more uniform and smoother block, which reduced building labour requirements for laying by 20 per cent. Others argued that it was more difficult to get the blocks out of the hand-operated machines and the blocks were often slightly damaged. It therefore seems likely that laying time is greater with blocks produced by hand-operated machines than with vibrated blocks.

The blocks produced by the different techniques are not therefore completely homogeneous. Whether the somewhat lower and more erratic quality of the hand-machine-produced blocks is a serious objection to them depends on the requirements. The engineer at Machakos County Council was perfectly satisfied with the results, as suitable for low-cost single-storey housing and other buildings (including schools).

The quarrier in Nanyuki was also satisfied. Only in Nairobi, where multistorey houses abound, are these characteristics a decisive disadvantage. In the statistics therefore no allowance has been made for 'inferior' quality of blocks produced by hand-operated machines, though these quality differences must be borne in mind in assessing the results.

The vibrating characteristic had other implications. Obviously it required fuel costs which the hand-operated machine did not.[6] It also raised repair and maintenance requirements since the machine was vigorously vibrated, with the block, hundreds of times a day. Thus these machines either have to be built of extremely tough material and well put together, or they need considerable repairs and have a short life. Length of life estimates are almost invariably largely guesses but there was sufficient uniformity about these guesses to make them of interest.

All the replies relating to the hand-operated machine (category 1) (of which four were in continual and five intermittent use) said the machines could be maintained indefinitely. Seven of the locally produced stationary vibrating machines (2a) were given lives of between two and ten years, two replies were for an 'indefinite' life. The estimate for the specially adapted machine (2b) was four years. The four estimates for imported stationary vibrating machines (3) were very vague; viz. 'indefinite', 'no estimate, but 17 years old and still going well', 'longer than local' and 'over ten years'. For laying machines there were two replies, one indicating a life of eight or nine years and the other concerning a machine in only occasional use, 14 or 15 years. Taken together, the estimates suggest that whereas hand-operated machines lasted indefinitely, irrespective of whether they were operated intermittently or continuously, the vibrating machines were of limited durability. The durability of vibrating machines clearly depends on use; one firm suggested that the life of its machine would be halved if they operated a double shift. From the opinions expressed, it would also seem that the local machine generally has a shorter life than its imported counterpart.

Estimates of length of life of a machine should not be divorced from repair and maintenance costs, since the life of a machine can be prolonged by heavy maintenance. Estimates of repair costs are possibly the most unreliable of all for a number of reasons. Since it was impossible to check independently, this figure provided an opportunity to exaggerate the costs of the business: it was also easy for some spares to be forgotten, and ambiguity about whether labour required was or was not included also reduces the comparability of the figures.[7] One year's repair costs may not be typical; and one would expect repair costs to increase with age. A further problem arose from the replacement of moulds. Mould replacement has not been included in the repair estimates, apart from the high figure for two of the laying machines and one

TABLE 10.4 Examples of annual repair costs in individual cases, by category and age of machine

| Category of machine | | | | | | | | | |
|---|---|---|---|---|---|---|---|---|---|
| *1* | | *2a* | | *3a* | | *3b* | | *4* | |
| *Age (years)* | *Repair costs (E.A.s)* | *Age (years)* | *Repair costs (E.A.s)* | *Age (years)* | *Repair costs (E.A.s)* | *Age (years)* | *Repair costs (E.A.s)* | *Age (years)* | *Repair costs (E.A.s)* |
| 2 | 0 | 1 | 200[1] | 3+ | 1000 | 2 | 1000[2] | 3+ | 2000[1] |
| 4 | 0 | 1½ | 0[3] | 6 | 1000–2000 | 17 | 2000[1] | 3+ | 2000[1] |
| 5 | 200 | 1½ | 2400 | 6 | 3000–4000[4] | | | 5 | 23,750[4,5] |
| 7 | 25 | 2 | 2000–3000 | 8–9 | 300–400[3] | | | 13 | 31,250[4,6] |
| 12/13 | 100 | 2 | 3000[1] | | | | | 14 | 1000 |
| 15 | 100[7] | 3 | 1000 | | | | | | |
| 17 | 0 | 4 | 2000 | | | | | | |
| 50 | 25 | | | | | | | | |

[1] Spares only. [2] Labour also included. [3] Machine only in intermittent use. [4] Includes pattern replacement. [5] Spares 12,500/-, mechanic 5000/-, wear and tear of tools 6250/-. Daily cost supplied; assumed 250 repair days per annum. [6] Spares 20,000/-, mechanic 5000/-, wear and tear of tools 6250/-. [7] After five years.

of the others. Mould replacement required was greatest for the stationary vibrating machines, less for the laying machines and did not apply to the hand machines. While the manufacturer of the imported machine estimated that replacement would be required once a year, one of the firms suggested replacement every 60,000 blocks, which is likely to be two or three times a year.

Table 10.4 shows the repair cost estimates together with the age of the machine when the survey was taken. Hand machines require virtually no repairs whether used sporadically or continuously. For all other types repairs were substantial if the machines were used continuously, though they appear to be less if only used sporadically. Any tendency for repairs to increase with age is imperceptible in the figures, though since they were prepared on different bases this is not conclusive. The higher average age of the imported vibrating machines might be regarded as independent evidence for the view that they have longer lives. However, since local production of machines had only started four years before the survey, this was the maximum age for these machines.

It can be concluded that the hand-operated machines last almost indefinitely, and require no repairs. In contrast, the vibrating machines, being subject to continuous use, have limited lives and require regular and sizeable repairs. Repair requirements for the local stationary vibrating machines are similar to the imported machines, but the latter generally last a good deal longer.

Vibrating machines provide for machine-paced operation as compared with hand machines. The pace of the machine in no case entirely determined the pace of work as the machine could be slowed down; but the machine did impose a regular rhythm which was lacking in the hand-operated machine. This impression was suggested more by watching operations than by any statistics. With vibrating machine block-making, each person performed his function in regular time ready for the next step. In hand operations the whole thing was more like making mud pies, filling the mould, hitting it and then waiting to see if it collapsed or not, as children do on a beach, rather than the factory operation which the mechanised version suggested. Several of those interviewed commented that a more regular (and faster) mode of operation was achieved by the mechanised version. It should be noted, however, that 'machine pacing' applied much more to the stationary vibrating machines, and less to the laying machines.

### SCALE

The output obtainable from the different machines varied according to the speed of operation. The speed of operation was in turn dependent on supporting machinery (such as mixers), number of employees and efficiency of operation. Since the hand machines were less suitable for hollow blocks, for comparability only solid blocks are considered here.

The figures should be reduced by about one-third for stationary machines producing hollow blocks. It is assumed that each machine is operating a full single shift – eight or nine hours – which is what the machines were doing except for those in only intermittent use.

Table 10.5 shows the rate of output per shift reported by the operators. There were significant design differences among the laying machines, with three different manufacturers and each machine of a somewhat different design. These differences are reflected in the very wide variation in output rate reported; it is therefore almost impossible to generalise about this group. Within the other categories all the machines were of a similar design, despite differences in machine manufacturer.

The locally manufactured machine was basically a copy of the small imported machine. Differences between these two categories were largely in the method of manufacture and materials used, not basic design.

Apart from the laying machines (category 4), therefore, variation in rate of output within each category is not to be attributed to machine design. A maximum speed of operation is imposed by machine design but variations below this occur according to number of workers, organisation of activity etc. Despite these variations Table 10.5 permits some generalisations. On average the hand-operated machine produced a substantially lower rate of output than the vibrating machines, about half the 9 × 9 rate of production and a bit over a quarter of the 6 × 9 rate of the locally manufactured vibrating machine. The locally manufactured and small imported stationary machines had a similar rate of output, as was to be expected given the basic similarity of design, while the rate of output of the large imported machine was over twice as great. The rate of output of three of the five laying machines was similar to that of the large stationary machines. The modified local machine's output was the same as the maximum rate reported among the unmodified machines.

In terms of *scale* of output the machines fall into three groups:

I the hand-operated (category (1));

II the small vibrating stationary, whether imported or locally produced (2a, 2b and 3a) and the two small laying machines (4a);

III the large vibrating stationary (3b) and the large laying (4b).

Table 10.6 shows output and employment associated with each machine according to scale. With the exception of the local modified machine, average employment requirements rise with the scale of output but less than proportionately, so that labour productivity rises, and labour input per block manufactured falls as the scale of output increases.

As can be seen from this data, the performance of the locally manufactured and the imported small stationary machine is very similar. However, within each category there are divergences in performance

TABLE 10.5 Number of blocks produced per shift of eight or nine hours in individual cases, by category of machine and size of block (averages in italics)

| *Category of machine* | | | | | | | | | |
|---|---|---|---|---|---|---|---|---|---|
| *1* | *2a* | | *2b* | *3a* | | *3b* | | *4* | |
| *Size of block* | | | | | | | | | |
| *9 × 9 in.* | *6 × 9 in.* | *9 × 9 in.* | *6 × 9 in. only* | *6 × 9 in.* | *9 × 9 in.* | *6 × 9 in.* | *9 × 9 in.* | *6 × 9 in.* | *9 × 9 in.* |
| 60–70 | 180[1] | 90[1] | 1800 | — | 800 | 2000 | 2000[2] | 525 | 350 |
| 150 < | 200 | 150 | | 500 | 300 | 2800 | 1400 | 875 | 525 |
| 200 | 600 | 400 | | 1200 | 600–750 | 3000 | 1200 | 2000 | 1250 |
| 200 | 1000 | 500 | | | | 4000 | 2000 | 2300[3] | |
| 200–300 | 1200 | 600 | | 1200 | 800 | 4000 | 2000 | 3400 | |
| 230 | 1200 | 700–800 | | | | | | | |
| 250 | 1600 | 800 | | | | | | | |
| 300 | 1800 | 1000 | | | | | | | |
| 300–400 | | | | | | | | | |
| >350 | | | | | | | | | |
| 500[4] | | | | | | | | | |
| *260* | *972* | *536* | *1800* | *967* | *644* | *3160* | *1720* | *1820* | |

[1] Used a different local machine from all others; also few workers and no mixing machines.
[2] Sometimes too fast for workers, then about 1750 at slower speed. This is a different type of imported machine that makes the same number of 6 × 9 in. or 9 × 9 in. blocks.
[3] Hollow, but makes little difference with laying machine.
[4] Flat out.

TABLE 10.6 Average output and employment, by scale of production and category of machine

| *Group and category of machine* | *Size of block (in.)* | *Output* (*O*) | *Average employment* (*L*) | *Blocks per man-day* (*O/L*) | *Man-days per block* (*L/O*) |
|---|---|---|---|---|---|
| Group I: | | | | | |
| 1 | 6 × 9 | 260 | 6 | 43·3 | ·023 |
| Group II: | | | | | |
| 2a | 6 × 9 | 948 | 8·5 | 111·5 | ·009 |
| | 9 × 9 | 536 | | 63·1 | ·016 |
| 2b | 6 × 9 | 1800 | 19 | 94·7 | ·011 |
| 3a | 6 × 9 | 967 | 9 | 107·4 | ·009 |
| | 9 × 9 | 644 | | 71·6 | ·014 |
| 4a | 6 × 9 | 700 | 9 | 77·8 | ·013 |
| | 9 × 9 | 438 | | 48·7 | ·021 |
| Group III: | | | | | |
| 3b | 6 × 9 | 3160 | 11·5 | 274·8 | ·004 |
| | 9 × 9 | 1720 | | 149·6 | ·007 |
| 4b | 6 × 9 | 2567 | 15·5 | 165·6 | ·006 |

according to the efficiency with which the machines are run, and the averages conceal considerable variations in output and employment. Averages are used there because if other 'typical' or 'representative' figures are used it is very easy to bias the evidence in selection. However, particularly for categories 2a and 3a, the average rate of output does seem to diverge from what appeared to be typical of a fairly efficient operation. In Table 10.7 two sets of figures have been used to represent this category of machine – the average for the group as a whole, and what is taken to be a 'typical' performance. The figures taken as 'typical' for categories 2a and 3a were rates of output of 1200 (6 × 9) and 800 (9 × 9) and employment of 12, compared with average figures of 950 (6 × 9) and 570 (9 × 9) and employment of 8·5. On the whole, labour requirements per block tend to fall with scale as labour productivity rises, so that the largest machines use about one-quarter (depending on which block is produced) of the labour of the hand machines, while the small power-driven machines use about half as much labour per block; labour intensity defined in terms of man-days per block therefore falls considerably as the scale of production rises. The labour involved was generally unskilled. Training on the job normally takes less than a day. Some of the firms had a foreman whom they paid above the unskilled rate, but he too was trained on site in a short period. Skills were required for repair and maintenance; as repair and maintenance requirements rose with scale, so did the requirements for skill.

TABLE 10.7 Performance of various categories of machine in relation to that of hand-operated machines (hand-operated machines = 1)

| *Group and category of machine* | *Size of block (in.)* | *Output* O | *Average employment* L | *Blocks per man-day* O/L | *Man-days per block* L/O |
|---|---|---|---|---|---|
| *Scale II:* | | | | | |
| 'Typical' 2a and 3a | 6 × 9 | 4·6 | 2·0 | 2·3 | 0·43 |
| | 9 × 9 | 3·1 | | 1·5 | 0·65 |
| Average 2a and 3a | 6 × 9 | 3·7 | 1·4 | 2·6 | 0·39 |
| | 9 × 9 | 2·2 | | 1·5 | 0·65 |
| 2b | 6 × 9 | 6·9 | 3·2 | 2·2 | 0·46 |
| 4a | 6 × 9 | 2·7 | 1·5 | 1·8 | 0·56 |
| | 9 × 9 | 1·7 | | 1·1 | 0·89 |
| *Scale III:* | | | | | |
| 3b | 6 × 9 | 12·1 | 1·9 | 6·3 | 0·16 |
| | 9 × 9 | 6·6 | | 3·5 | 0·29 |
| 4b | 9 × 9 | 9·9 | 2·6 | 3·8 | 0·26 |

INVESTMENT AND RUNNING COSTS

The firms supplied figures for the cost of acquisition of the machines. Since they were bought at varying times in the past, with changing prices, the figures supplied may not be on a comparable basis; moreover, some of the machines were bought new, others second-hand, and some firms gave a gross figure only, inclusive of a mixer and other capital costs. Table 10.8 shows the acquisition cost, as reported with no adjustments. Table 10.9 shows the investment per man employed and investment productivity, or daily output per shilling of acquisition cost, using the unadjusted acquisition cost. The substantial variations between machines in each group make it difficult to generalise. On average, investment costs per worker rise as the scale of the machine increases. The locally produced vibrating machines have an acquisition cost per worker, on average, nearly three times that of the hand machines. The small imported machines involve nearly twice as much investment per worker as the locally produced machines, while the large imported stationary machines require eight times the investment per man of the hand machines and over one-and-a-half times that of the small imported machines. The investment-output ratios in Table 10.10 are for 6 × 9 in. blocks. If the figures for 9 × 9 in. blocks were used, the ratio for all categories except hand machines would roughly double, and hence the comparison would be substantially more favourable to the hand machines.

The ratios used are not comparable to capital-output ratios as normally defined – that is the ratios of capital stock to annual output in value terms: the $(I/O)$ ratios are ratios of acquisition cost to daily output in volume terms. The unadjusted ratios suggest that, on average, the investment-output ratio for hand machines is lower, and investment productivity higher, for hand than for mechanised machines. Thus the hand machines appear technically efficient, saving investible resources in relation to output, as well as employment. On the other hand, the ratios show greater investment costs in relation to output for the small vibrating machines (whether imported or locally produced) than for the large vibrating or laying machines, suggesting that the small vibrating machines may be inefficient at any factor prices, using more of both factors as compared with the large machines.

TABLE 10.8 Acquisition cost of machines (E.A.*s*; second-hand purchases in italics)

| *Category of machine* | | | | | |
|---|---|---|---|---|---|
| *1* | *2a* | *2b* | *3a* | *3b* | *4* |
| *0* | *6000* | 24,000[1] | *6000* | *15,000–20,000* | *6000* |
| *500* | 7000[2] | | 23,000 | 35,000 | *13,000* |
| *1000* | 10,000 | | *25,000–30,000* | 35,000 | 30,000 |
| 1800 | 10,000 | | 28,000 | *60,000* | 400,000[3] |
| 1800 | 10,700 | | | *110,000*[1] | 500,000[3] |
| 2000 | 12,000 | | | | |
| 2800 | 12,000 | | | | |
| 4000 | 12,000 | | | | |
| 5000 | | | | | |
| 8000 | | | | | |

[1] Includes mixer, chutes, etc.
[2] Cost to make.
[3] Estimate of capital cost of full plant if new 1969. Estimated cost of similar block-making machine only, 54,000/-.

Historic acquisition costs may be of little relevance to current opportunities. For the economy as a whole current opportunities include making or importing new machines or importing second-hand machines. None of the data included imported second-hand machines. Surveys of producers and retailers of equipment together with the information supplied by users who had recently (i.e. within a year or so of the survey) bought equipment new, suggested the prices indicated in Table 10.10, which also shows the various ratios calculated on the basis of these prices. Investment intensity, when defined as investment per man, is substantially higher for all the power-driven categories than for the

TABLE 10.9 Individual and average ratios of acquisition costs[1], employment[2], and output[3] (6 × 9 in. blocks)

| Category of machine | | | | | | | | | | | |
|---|---|---|---|---|---|---|---|---|---|---|---|
| *1* | | *2a* | | *2b* | | *3a* | | *3b* | | *4* | |
| $\frac{I}{L}$ | $\frac{I}{O}$ | $\frac{I}{L}$ | $\frac{I}{O}$ | $\frac{I}{L}$ | $\frac{I}{O}$ | $\frac{I}{L}$ | $\frac{I}{O}$ | $\frac{I}{L}$ | $\frac{I}{O}$ | $\frac{I}{L}$ | $\frac{I}{O}$ |
| 100 | 2·5 | 333 | 3·3 | 1263 | 13·3 | 1833 | 34·4[4] | 2188 | 5·8 | 667 | 11·4 |
| 200 | 15·3 | 892 | 7·6 | | | 2000 | 12·0 | 2917 | 8·75 | 1444 | 14·9 |
| 286 | 5·7 | 1273 | 43·7 | | | 2333 | 23·3 | 3273 | 12·5 | 2143 | 15·0 |
| 300 | 7·2 | 1412 | 10·0 | | | 3833 | 19·2 | 5435 | 30·0 | 3000[6] | 23·5[6] |
| 300 | 7·8 | 1600 | 12·0 | | | | | 6875[5] | 27·5[5] | 3600[6] | 15·9[6] |
| 467 | 11·2 | 1667 | 8·3 | | | | | | | | |
| 500 | 17·8 | 2000 | 20·0 | | | | | | | | |
| 1000 | 16·7 | 2000 | 50·0 | | | | | | | | |
| 1333 | 16·0 | | | | | | | | | | |
| *498*[7] | *11·1*[8] | *1397* | *19·4* | *1263* | *13·3* | *2500* | *22·2* | *4137* | *16·9* | *2171* | *16·1* |

Figures in italics denote averages. Key: $I$ = investment; $L$ = Labour; $O$ = Output.

[1] In East African shillings.

[2] In man-days.

[3] Number of blocks produced per day.

[4] 9 × 9 in.

[5] Includes mixer and chutes.

[6] Using estimated cost for block-maker alone derived from costs for similar machine.

[7] The averages are 598 for machines bought new and 150 for those bought second-hand.

[8] The averages are 11.8 for machines bought new and 9.0 for those bought second-hand.

TABLE 10.10 Investment/labour, investment/output and output/labour ratios by category of machine (6 × 9 in. blocks)

| *Category of machine* | E.A. '000*s* | *Daily output* | *Employment* | $\frac{I}{L}$ | $\frac{I}{O}$ | $\frac{O}{L}$ |
|---|---|---|---|---|---|---|
| 1 | 2[1] | 260 | 6 | 333 | 7·7 | 43·3 |
| 2a 'Typical' | 12 | 1200 | 12 | 1000 | 10·0 | 100 |
| Average | | 950 | 8·5 | 1412 | 12·6 | 111·8 |
| 2b | 24[2] | 1800 | 19 | 1263 | 13·3 | 94·7 |
| 3a 'Typical' | 23 | 1200 | 12 | 1917 | 19·2 | 100 |
| Average | | 950 | 8·5 | 2706 | 24·2 | 111·8 |
| 3b | 35 | 3160 | 11·5 | 3043 | 11·1 | 274·8 |
| 4b | 54[3] | 2300 | 15 | 3600 | 23·5 | 153·3 |

[1] Price of locally produced machine; import price 2100/-.
[2] Includes chutes, mixers, etc.
[3] Within the laying category (4), reliable data including acquisition cost for a new machine were available only for one laying machine, with a daily output rate of 2300 6 × 9 in blocks, so that only the data for this machine are included in the table.

hand-operated machine. Investment requirements per unit of output are also lowest for the hand-operated machines, while labour requirements per unit of output are highest for the hand machines. Table 10.11 shows the ratios expressed as a proportion of the figures for the hand-operated machines.

TABLE 10.11 Investment/labour, investment/output[1] and labour/output ratios of various categories of machine in relation to those of hand-operated machines (hand-operated machines = 1)

| *Category of machine* | *Cost of machine* | $\frac{I}{L}$ | $\frac{I}{O}$ | $\frac{O}{L}$ | $\frac{L}{O}$ |
|---|---|---|---|---|---|
| 1 | 1·0 | 1·0 | 1·0 | 1·0 | 1·0 |
| 2a 'Typical' | 6·0 | 3·0 | 1·3 | 2·3 | 0·43 |
| Average | | 4·2 | 1·6 | 2·6 | 0·38 |
| 2b | 12·0 | 3·8 | 1·7 | 2·2 | 0·45 |
| 3a 'Typical' | 11·5 | 5·8 | 2·5 | 2·3 | 0·43 |
| Average | | 8·1 | 3·1 | 2·6 | 0·38 |
| 3b | 17·5 | 9·1 | 1·4 | 6·4 | 0·16 |
| 4b | 27·0 | 10·8 | 3·1 | 3·5 | 0·29 |

[1] Output in 6 × 9 in. blocks.

Transforming the initial investment cost into an annual equivalent,[8] to take into account length of life, makes quite a difference to the comparison, as Table 10.12 shows. Because of its very long life the hand machine appears substantially cheaper than before, while the short life of the locally produced vibrating machine just offsets its lower acquisition cost as compared with the imported machine. At a discount rate of 10 per cent the annual investment costs of the two types of small stationary machine (locally produced and imported) become very similar. The extremely short life assumed for the specially adapted local machine substantially increases its relative investment costs, so that at low rates of discount it exceeds that of all other techniques.

TABLE 10.12 Annual investment cost by category of machine (E.A.*s*)

| *Category of machine* | *Estimated life (in years)* | *Annual investment cost at percentage discount rate indicated* | | |
|---|---|---|---|---|
| | | *0* | *10* | *20* |
| 1 | 30 | 70 | 210 | 400 |
| 2a | 5 | 2400 | 3170 | 4010 |
| 2b | 4 | 6000 | 7570 | 8030 |
| 3a | 15 | 1530 | 3030 | 4920 |
| 3b | 15 | 2340 | 4600 | 7490 |
| 4b | 12 | 4500 | 7930 | 12,170 |

The annual adjustment (as shown in Table 10.13) improves the relative performance of the hand machine for both ratios: the relative improvement is greater the lower the interest rate. The ordering of the investment/labour ratio is not seriously affected by the annual adjustment, although the small stationary machines come much closer together, while the relative *I/L* of the adapted local machine increases substantially.

It remains true, broadly,[9] that the investment/labour ratio rises as the scale increases. All techniques use more investment in relation to output than the hand techniques: how much more depends on the exact basis of the calculations.

When the mechanically-powered techniques are compared, the large imported stationary machine is found to have significantly the lowest investment/output ratio: it also has the highest output/labour ratio. In contrast to this, the laying machine and the small stationary vibrating machines appear to represent inferior techniques, since their investment and labour costs per unit of output are both higher than those of 3b.

TABLE 10.13

| *Category of machine* | *Investment/labour ratio* | | | *Investment/output ratio* | | |
|---|---|---|---|---|---|---|
| | *Unadjusted acquisition cost* | *Annual cost discounted at* | | *Unadjusted acquisition cost* | *Annual cost discounted at* | |
| | | 0% | 20% | | 0% | 20% |
| 1 | 1·0 | 1·0 | 1·0 | 1·0 | 1·0 | 1·0 |
| 2a 'Typical' | 3·0 | 17·9 | 5·0 | 1·3 | 7·7 | 2·2 |
| Average | 4·2 | 25·3 | 7·0 | 1·6 | 9·7 | 2·7 |
| 2b | 3·8 | 28·3 | 6·3 | 1·7 | 12·8 | 2·9 |
| 3a 'Typical' | 5·8 | 11·4 | 6·1 | 2·5 | 4·9 | 2·6 |
| Average | 8·1 | 16·2 | 8·6 | 3·1 | 6·2 | 3·3 |
| 3b | 9·1 | 18·2 | 9·7 | 1·4 | 2·8 | 1·5 |
| 4b | 10·8 | 26·8 | 12·1 | 3·1 | 7·5 | 3·4 |

This is illustrated in Figure 10.1. All the techniques are to the north-east side of the line joining the hand technique to the large imported stationary machine. This indicates their inferiority. This is true both at 0 and at 20 per cent interest. The diagram does not take the scale of production into consideration. Since the techniques are designed for different scales the conclusions may be altered when scale is considered. There are also important dimensions of cost which have been omitted from the

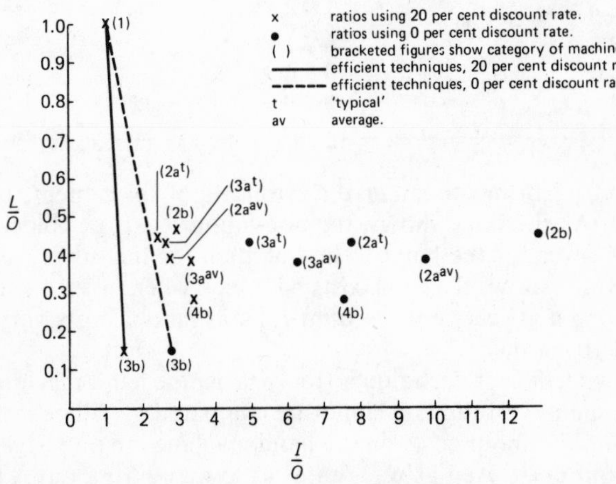


Fig. 10.1 Relative efficiency of block-making techniques using different types of machine

analysis. These are repair costs, mould replacement and fuel consumption.

As seen earlier, the different techniques are associated with different annual repair costs. The vibrating machines also require periodic replacement of patterns or moulds in which the blocks are formed. This does not apply to the hand machine. Mould replacement costs between 2000 and 4000/- a year for the small stationary vibrating machines and somewhat more for the larger machines which require more expensive moulds. Allowing for pattern replacement and repairs would raise the annual costs of the vibrating techniques by between 4000 and 8000/- a year, which is as much as or more than the annual adjusted investment costs. The repair costs of the hand machines were much lower, about 50/- a year. Inclusion of repair costs and mould replacement reduces the relative costs of the hand machine, but does not significantly alter the comparison between the stationary vibrating machines.

Estimates for fuel consumption were supplied on a number of different bases. For some no estimates were supplied; for others a total figure was given inclusive of mixer, and sometimes quarry works. Piecing the evidence together, and deducting for quarry and mixer, suggests the approximate estimate shown in Table 10.14.

TABLE 10.14 Fuel costs by category of machine

| *Category of machine* | *Per block*, (cents) | *Per annum*, (E.A.*s*) |
|---|---|---|
| 1 | 0 | 0 |
| 2a} 3a} | 1-2·5 | 2500-6250 |
| 3b | 0·7-1 | 5250-7500 |
| 4a | 0·3-0·9 | 3060-3150 |
| 4b | 0·3-0·9 | 5175 |

Table 10.15 brings together the estimates of investment, repair and fuel costs. As the table shows, the non-labour costs per block are substantially lower for the hand technique than for the other techniques.

Table 10.16 shows the total costs per block when investment costs are valued using a 10 per cent discount rate, as labour costs vary between 2·5/- and 10/- a day.

If the two 'efficient' techniques (those using the hand machine and the large stationary vibrating machine) are compared, it will be seen that the savings in non-labour costs on the hand machine are outweighed by the extra labour costs even at wage-rates as low as 2·5/- a day. (The minimum urban wage at the time was 7/- a day; the estimated opportunity costs of rural labour[10] around 2 or 3/- a day). With a 10 per cent discount

TABLE 10.15 Annual non-labour costs by category of machine (E.A.*s*)

| *Type of cost* | *Category of machine* 1 | 2a[1] 3a | 2b | 3b | 4b |
|---|---|---|---|---|---|
| Annual investment[2] | 210 | 3100 | 7570 | 4600 | 7930 |
| Repairs and moulds | 50 | 5000 | 10,000[4] | 8000 | 15,000 |
| Fuel[3] | 0 | 4380 | | 6380 | 5180 |
| Total[5]: | 260 | 12,480 | 17,570 | 18,980 | 28,110 |
| *Cents* per block | *0·4* | *5·3* | *3·9* | *2·4* | *4·9* |
| *Ratio* (hand operated machine = 1) | *1* | *13·1* | *9·8* | *6·0* | *12·2* |

[1] Estimates for the imported and locally produced stationary vibrating machine are so close that they are treated together here.
[2] Assuming a discount rate of 10 per cent.
[3] Taking mid-point of fuel estimates shown in Table 10.14
[4] Estimate.
[5] Assuming 250 working days a year.

TABLE 10.16 Costs per block by category of machine (E.A. cents)

| *Type of cost* | *Wage rate* (E.A.*s*) | *Category of machine* *1* | *2a, 3a* | *3b* | *3b* | *4b* |
|---|---|---|---|---|---|---|
| Non-labour[1] | | 0·4 | 5·3 | 3·9 | 2·4 | 4·9 |
| Labour, at wage rates indicated | 2·50 | 5·8 | 2·2 | 2·7 | 0·9 | 1·6 |
| | 5·00 | 11·6 | 4·5 | 5·3 | 1·8 | 3·3 |
| | 7·00 | 16·2 | 6·2 | 7·4 | 2·5 | 4·6 |
| | 10·00 | 23·1 | 8·9 | 10·6 | 3·6 | 6·5 |
| Total, at wage rates indicated | 2·50 | 6·2 | 7·5 | 6·6 | 3·3 | 6·5 |
| | 5·00 | 12·0 | 9·8 | 9·2 | 4·2 | 8·2 |
| | 7·00 | 16·6 | 11·5 | 11·3 | 4·9 | 9·5 |
| | 10·00 | 23·5 | 14·2 | 14·5 | 6·0 | 11·4 |

[1] Assuming a discount rate of 10 per cent.

rate the switching wage for these two techniques is 1/03 – i.e. at wages above this the vibrating machine is cheapest, while at wages below it the hand machine is. Using discount rates below 10 per cent would raise the switching wage, while using discount rates above 10 per cent would reduce it. The costs per block of the other powered techniques are substantially higher than that of the large stationary vibrating machine. The small stationary vibrating machine is nearest to the hand machine in price and scale, and consequently in some sense its closest competitor. The switching wage between these two techniques, again using a 10 per cent discount rate is 3/40, which is higher than the competitive rural wage rate, and probably higher than the urban opportunity cost of labour. Irrespective of whether it is locally manufactured or imported, the small stationary vibrating machine is a product of an 'old' technology from an advanced country; and it is accordingly interesting to compare its costs with those of the technique especially adapted to local conditions. In fact, it appears that total costs were very similar, with the adapted technique involving a somewhat higher element of labour costs and lower non-labour costs. Hence the adaptation, though not dramatic, was in the right direction.

The calculations have so far ignored scale. In fact each of the techniques is indivisible in the sense that investment costs cannot be saved by operating at a lower scale. If operations were on a smaller scale the other costs would be less but not proportionately less. If we assume that investment costs are invariant, with respect to output, but that all other costs vary proportionately with output, then if all techniques were utilised to the point at which output was equal to the eight-hour output of the hand techniques (working 250 days a year), the investment costs per block of all the techniques other than the hand technique would rise correspondingly. The switching wage between the hand machine and the large stationary vibrating machine would then be 4/35. In the rural areas, where wages are generally less than this, the hand machine would be the sensible choice. In the urban areas, the social cost of labour is probably lower than this figure.

These calculations assume that all non-acquisition costs are variable. In fact there are other costs that are unlikely to vary proportionately with output. Suppose we assume that labour costs vary proportionately with output but that half of the repair, maintenance and fuel costs do not vary with respect to output, then to use the powered techniques to produce the level of output of the hand machine would raise their costs significantly more. For example, the switching wage between hand machines and large stationary vibrating machines would rise to just over 9/-. In this situation the costs of the small stationary vibrating machine would be less than those of the large machine for wages of less than 9/43.[11]

The figures are illustrative of the importance of scale and capacity utilisation in determining the relative costs of different techniques designed for different scales of output. Because of heavy transport costs (a solid 6 × 9 × 9 in. block weighs 52 kg.) the scale of production is determined by the market for blocks in the immediate vicinity. In rural areas the market is often smaller than the output obtained from eight-hour capacity use of a hand machine, and the machines lie idle much of the time. In such situations, while the costs of all the techniques would rise, those of the hand machine would rise least, since the greatest proportion of its costs are labour and therefore escapable; conversely, if each of the techniques were operated for more than one shift a day all costs would fall, but those of the investment-intensive techniques would fall most.

The considerations discussed suggest why rural and urban choice of technique may differ. First, there is the question of product standard; blocks in large cities may have to be strong enough for more than one-storey accommodation. Secondly, there is the question of scale. Thirdly, wage costs differ, wage rates being lower in rural areas. All these considerations indicate that more labour-intensive hand-operated machines are more suitable for rural use.

## THE PROCESS APPROACH

The production of any good can be split into a number of activities. The analysis so far has been entirely concerned with examining the costs associated with different types of block-making machinery. But in the production of blocks the operation of the actual block-making machine is only one part of the production process. The production process as a whole can be split up as follows:

1. Production of raw materials.
2. (*a*) Transfer to site.
   (*b*) Transfer to mixing area.
3. Mixing raw materials.
4. Transfer of mixture to block-making machine.
5. Operation of block-maker.
6. Transfer of blocks from machine to drying area.
7. Watering blocks when drying.
8. Stacking blocks.
9. Transporting blocks from site to where needed.

The first phase itself covers a large number of activities. The divisions are to some extent arbitrary, and reflect the type of machines available. Thus, if a single machine were always used to mix and produce the blocks these two activities might be amalgamated into one. In each of the activities described some choice of technique is possible. Thus transport (of materials and then of blocks) may be done in a number of ways, depending on distance, quantities and costs. The technique used in the production of raw materials was not included in the survey.[12]

The following range of techniques was observed for different activities:

2(*a*) and 9. Transport to and from site: lorry.
2(*b*). Transfer to mixing area: spade, wheelbarrow or lorry, depending on scale and distance.
3. Mixing: spade or either local or imported mixing machine (diesel or electric).
4. Transfer of mixture to block-making machine: spade or automatic chute.
5. Block-maker, already covered.
6. Transport of blocks to drying area: manual or by wheeled trolley.
7. Watering: women with watering cans or hoses or automatic hoses.
8. Stacking: manual or by hand-operated trolley or electrically operated stacking machines.

There was no unbreakable connection between choice of technique at one stage and choice at another: in theory one could combine very labour-intensive mixing (on the ground) with a more mechanised block-maker, but in practice, choice of technique at one stage did partly determine choice at another. This was in part a question of speed, and in part of scale. The large block-makers required a steady and rapid flow of mixture – which virtually ruled out mixing on the ground; generally the larger machines were combined with automatic mixers so that the mixture was automatically transferred to the machine. The quality of the blocks, in terms of strength and uniformity for a given ratio of materials, tended to be greater for automatic than for manual mixing, and also, it was claimed, if the mixture was transferred automatically from mixer to machine.

In general, therefore, the most labour-intensive methods of block-making (the hand machines) were also combined with the most labour-intensive methods of mixing, etc., while the converse applied for the more investment-intensive methods. This partly reflected different attitudes to mechanisation on the part of the entrepreneurs, which influenced all their decisions; but it also reflected the fact that requirements of scale and standard of block manufacture which led to the adoption of investment-intensive methods of block manufacture led to similar decisions in mixing. For stacking and watering the use of mechanised methods was confined to the larger mechanised block-makers.

Since in practice the type of block-maker does partly determine the mixing method, a comparison of costs of techniques should take this into account. In the typical case the hand machines were used with manual mixing, the small vibrating machines with small local or imported mixers and the larger vibrating machines with larger mixers combined with a system of automatic chutes. The typical investment costs might therefore be affected as shown in Table 10.17. Inclusion of costs of mixing may thus make a substantial difference to the comparison. Very roughly the cost of the typical mixer is in line with the cost of the block maker. Mixers also need repairs and use fuel. Generally, the repair cost seemed substantially less than that for the block-makers (about one-third), but the fuel costs were of a similar magnitude. On the assumption that the hand-operated machines use no mixer and that their investment costs are unchanged but that the non-labour costs of the other categories increase by 50 per cent as compared with earlier estimates (Table 10.16), the costs per block become those shown in Table 10.18. Inclusion of the mixer increases the costs of the mechanised machines in relation to the hand machines but does not alter the overall results.

Other investment costs might also be included. The costs of erecting the necessary buildings and floor can be substantial, but since such costs are not associated with particular block-making machines they have not

TABLE 10.17 Typical investment costs of cement-mixing techniques, by type of equipment

| *Type of mixing equipment* | *Cost* (E.A.*s*) | *Suitable for use in association with* |
|---|---|---|
| Spade | 25 | (1) hand-operated block-maker<br>(2) possibly small vibrating machine |
| Small mixer | 6000-10,000 | (1) hand-operated block-maker<br>(2) small vibrating machine |
| Large mixer | 40,000-50,000 | (1) large stationary vibrating machines<br>(2) laying machines (large) |

TABLE 10.18 Costs per block by category of machine, including provision for corresponding mixers (E.A. cents)

| *Category of machine* | *Size of block (in.)* | *Non-labour costs*[1] | *Total costs at wage rate indicated* (E.A.*s*) | |
|---|---|---|---|---|
| | | | 5·00 | 10·00 |
| 1 | | 0·4 | 12·00 | 23·50 |
| 2a and 3a | 6×9 | 8·2 | 12·90 | 17·10 |
| 3b | 9×9 | 3·6 | 5·40 | 7·20 |
| 4b | 6×9 | 7·4 | 10·70 | 13·90 |

[1] Discounted at 10 per cent, and taken from Table 10.16: category 1 same as in Table 10.16; other categories one-and-a-half times the figures in Table 10.16.

been included. Generally, inclusion of these costs is likely to raise the costs of all the powered techniques, in relation to the use of the hand machine, since the latter is often used with a minimum of infrastructural and building investment.

PRODUCT CHOICE: OTHER BUILDING MATERIALS

The survey covered different methods of making cement blocks in Kenya. However, the labour-intensity of material manufacture and construction may also be altered by selecting different building materials. To supplement the survey, therefore, some of the other materials available in Kenya are briefly described below, with a report on the associated input requirements based on interviews conducted in 1972.

*Mud and wattle.* These are the materials from which traditional housing is made; the vast majority of the rural population are housed in mud and wattle houses. The construction of these houses is almost exclusively labour-using, in the sense that they are made with local materials, local labour and no purchased equipment.[13] They are not suitable for much urban housing because of lack of materials within easy distance, and because they generally fall below the standards set by urban housing authorities. However, although they do not meet those requirements, they are often of far better quality than the houses made of cardboard boxes, newspapers, etc., which many urban dwellers use as building materials. At Thika, the City Council has prepared some experimental houses with mud and wattle walls covered with various plaster and cement washes. The Council found these unsatisfactory, as they deteriorated rapidly and required more labour and materials than dried blocks. Mud and wattle is labour-intensive in terms of maintenance requirements as well as of initial construction. It is therefore most suitable, and most often to be found, as a building method where labour is costed at close to zero, as in the subsistence sector.

*Sun-dried clay blocks.* These are made entirely from local materials. The clay is mixed with *murram*[14] and a little grass, and moulded in wooden moulds; the blocks are then dried in the sun. It is estimated that one man makes up to ten blocks in a day (not including the labour for digging up the materials). Block manufacture thus normally includes no purchased equipment, and in this sense is almost purely labour-intensive.[15] Labour used in constructing buildings made of these blocks is similar to that in concrete block buildings. The walls may be covered with cement or plaster to improve their resistance to the rain. The production of clay blocks is more labour-intensive than that of concrete blocks, both in the manufacture of materials to make the blocks and in the manufacture of the blocks themselves; but the quality of the blocks is inferior. This type of block can be produced in a block-maker similar to that used for concrete blocks. In one case where a machine was used, six untrained men produced 450 blocks in an hour, or nearly

80 per man per day. The machine cost 1500/- and was similar to the hand-operated block-makers described above.

A stronger clay block was made by the prisoners at Thika. To form the blocks they used a machine (costing 50,000/- in 1956), which made 2500 blocks a day and employed 100 people, including those digging and moving the blocks around. It is impossible to compare the (apparently low) labour productivity with that of the concrete block-makers because of the inclusion of labour for digging in the prison figures. The investment/output ratio (at 20) was similar to that of the mechanically operated block-makers and above that of the hand machine (see Table 10.10). The very low labour cost (the prisoners were paid 10–12 cents a day) was passed on in the block price, which as a result was also very low.

*Murram-enforced blocks*. These are made from murram, cement and sand. On the Nairobi City Council estate they were made with a hand-operated block-maker, which cost 1800/- when new. Four men could produce 300 blocks a day. The chief difference between this operation and that of hand-operated cement block manufacture was the use of *murram* instead of stones. This led to a considerable cost saving – a typical block cost 40 cents instead of 1/25 – but the quality was substantially lower. The crushing strength of the murram blocks was found to be 150–220 lb./sq. in. on one testing, compared with the official requirement of 400 lb./sq. in. The labour requirements for building were slightly above those for concrete blocks because of the uneven quality of the blocks.

*Black cotton bricks*. Black cotton is the type of soil found in many areas where the richer murram is absent. It is mixed with water and put in moulds. After drying and removal from the moulds the bricks are stacked and dried, using a primitive oven consisting of a hole in the ground, as in traditional charcoal burning. Two people can produce 250 bricks a day. (The bricks are about half the size of a normal cement block.) Black cotton bricks are not as strong as murram but are cheaper in the areas where murram is not available.

*Stone*. The possibility of using stone depends on the availability of nearby quarries since stone is too heavy to transport far. The use of stone is highly labour-intensive. It is estimated that one man can quarry six blocks of stone in one (intensive) hour. They then require a good deal of skilled or semi-skilled labour for building because of their uneven edges. The resulting building is of high quality.

*Timber*. Building with timber tends to be fairly skill-intensive, though little capital equipment is required for either production or building. In an experimental construction project,[16] Nairobi City Council employed nearly as many carpenters (15) as unskilled labourers (20) on the site, and considerably more carpenters (eight to ten) than labourers (about four) in the workshop where the panels were prepared.

A breakdown of the costs of a single-storey timber house (total costs per house E.A.£800–850) showed that labour costs varied between 30 and 47 per cent of total costs, material costs varied between 47 and 53 per cent and plant cost between 4 and 11 per cent. The proportion of labour costs was higher than that for the construction sector as a whole in 1962;[17] labour costs accounted for 29·2 per cent of gross output, material inputs for 51·3 per cent and operating surplus for 19·5 per cent. Three different low-cost housing projects around Nairobi gave labour costs as 30 per cent of total costs, with materials accounting for between 50 and 60 per cent.[18] These three schemes used cement blocks or reinforced murram blocks. The relatively high proportion of labour costs in the timber project reflects high skill levels and wage rates rather than greater employment generation. Moreover, a greater proportion of labour employment in preparation of materials has been included, as compared with the other figures.

*Pre-cast concrete panels.* There are various methods of forming pre-cast concrete panels. Broadly speaking, the manufacture is similar to that of the manufacture of cement blocks, and involves mixing cement and other materials, vibrating to strengthen and then leaving to dry. The labour-intensity of the method varies according to the machinery used, which tends to vary with scale. One pre-cast production unit visited had the following characteristics:

| | |
|---|---|
| Cost of mixer | 1500/- |
| Cost of air vibrator | 30,000/- |
| Cost of moulds | no estimate |
| Output | 14 panels a day (8 ft. × 3 ft.) $\simeq$ 300 blocks per day |
| Cost of wire for reinforcement | 5/- per panel |
| Employment | five unskilled labourers |
| Output/labour ratio (block equivalent) | 60 blocks per man-day |
| Investment/labour ratio ex-mixer and ex-moulds | 6000/- per worker |
| Investment/output ratio ex-mixer and ex-moulds | 100/- per block |
| Investment-labour ratio including mixer and allowing 2000/- for moulds | 6700/- per worker |
| Investment-output ratio including mixer and ex-moulds | 112/- per block |

Mixture used: one part cement, two of sand and four of ballast.

If this one example of pre-cast production unit is compared with the earlier figures for cement block manufacture, it appears that the ratio of investment cost to output and investment cost per worker is higher in the case of the pre-cast unit than in any examples of block manufacture. The rate of labour productivity, moreover, appears to be somewhat lower than that of most of the block-makers. However, this might be offset by improved labour productivity in construction, since the pre-cast units avoid much of the labour involved in laying blocks. But, the general impression given was that such labour saving in construction was slight because the panels were so heavy that a considerable amount of manpower was needed to lift and place them in position. Hence the pre-cast technology observed would appear to be inferior to the block-making technology, being more investment-intensive with little if any consequent labour saving.[19] This conclusion was supported by the fact that demand for the existing units was extremely low, and the factory was producing them only intermittently.

*Pre-cast concrete panels made with a foaming agent*. The major problem with pre-cast concrete panels is their weight. This problem has been partially overcome by a new technology involving the addition of a foaming agent to the mixture, which by creating bubbles in the panel improves insulation and makes it lighter. One production unit using such an agent was observed. It was pre-casting the panels on site for the building of 93 houses, and had the following characteristics:

| | |
|---|---|
| Cost of mixer from the United Kingdom | 29,000/- |
| Cost of motorised carrier | 18,000/- |
| Cost of moulds (100 at 500/- each) | 50,000/- |
| Output | 50 panels (of 2·3 square metres a day) $\simeq$ 1000 blocks a day |
| Employment | 30 workers (of whom two skilled) |
| Output/labour ratio (block equivalent) | 33·3 blocks per man-day |
| Investment/labour ratio ex-moulds and ex-carrier | 967/- per worker |
| Investment/output ratio ex-moulds and ex-carrier | 29/- per block |
| Investment/labour ratio including moulds and carrier | 3233/- per worker |
| Investment/output ratio including moulds and carrier | 97 shillings per block |

Mixture: one part cement and three of sand and foaming agent.

Labour productivity in the manufacture of these pre-cast panels appears to be somewhat lower than in the making of cement blocks. Investment per worker is higher than in the case of some of the cement block-making techniques. Much depends on how many moulds are included, what allowance is made for other equipment such as metal pipes, and whether costs of the mixer are included. Investment costs per unit of output tend to be higher than in the case of the block-makers for cement. In general, it seems that this type of production of panels is somewhat more investment-intensive than block-making. The cost of the foaming agent, owned and licensed by a United Kingdom company, must also be added. It is in construction, however, that the chief differences arise. The use of these pre-cast panels yields a considerable saving, which the manager of this unit estimated at 80 per cent, in labour. It also requires more capital equipment in construction – particularly cranes. Consequently, overall, the use of the panels involves greater investment-intensity and less labour use than concrete blocks. Although it would require much more detailed research to establish precise magnitudes, this conclusion emphasises the validity of the proposition discussed above: namely that looking at one stage of a chain production process can be highly misleading.

*Cement blocks made with a chemical additive.* Recent technological developments include the addition of a foaming agent to the mixture for the manufacture of cement blocks, making them lighter and better insulated. The operation requires careful supervision by scientifically qualified personnel, and is for large-scale production. The Ministry of Works received proposals for the establishment of a factory making these blocks from a multinational enterprise. The estimated capital cost of the proposed factory was E.A.£1 million, and the estimated labour force 150–200 workers and supervisors – a capital cost per work-place of between E.A.£5000 and E.A.£7000 or between E.A.100,000/- and E.A.140,000/- all told – over ten times the capital cost per worker of any of the block-makers. Among those employed 19 were to be managerial, including a number of chemists. Total block production was estimated at 400 cubic metres per year, on a two-shift basis, which is enough to build 7000 houses. Since the total number of houses built in Kenya in 1970–71 was 7000, the factory would displace most of the existing block-makers. There is an estimated 10 per cent saving in construction labour because of the even quality of the blocks. The very high quality of the blocks – even, light and strong, meeting all international specifications – was stressed. The material requirements are 5–10 per cent cement, 70 per cent lime and the balance sand. Particular specifications of lime and sand are required, so that the materials would need to be either transported over considerable distances or imported; to meet the requirements locally might well involve mechanisation of the production of materials to ensure the uniform quality required. The proposal for this factory, which had not (in 1973) been accepted, illustrates many aspects of choice of technique. First, the relationship of choice of technique to product specifications; given sufficiently detailed product specifications to meet international requirements, the proposed factory might seem the only possibility available, despite the investment-intensity of the technique. Secondly, technological developments in the advanced countries threaten to overtake older and more labour-intensive techniques. While in fact the proposal would probably have involved a higher-cost block, when further technical advances increased the factory's productivity it might have produced cheaper blocks than the ordinary cement block-makers, and consequently have displaced 2000 or so workers in block manufacture and as many more in construction. Thirdly, indivisibilities in modern techniques are such that marginal additions to productive capacity, which may meet the additional demand without creating excess capacity, are often not possible.

## CONCLUSIONS

If choice of building materials is included in choice of techniques, the choice is widened well beyond that which appears if only one building material is considered. Once a range of products is included, two aspects of choice of technique, easy to ignore in looking at a single material, become of key importance: one is the question of product quality, the other is the need to look at the productive process as a chain in which choice of technique at one stage helps to determine choice of technique at other stages. In relation to building materials the following steps in the chain need detailed examination:

(1) Production (excavation and processing) of raw materials.
(2) Transport of raw materials to next stage of processing.
(3) Processing of raw materials – i.e. block manufacture, preparation of timber panels, etc.
(4) Transport to the building site.
(5) Building techniques.

The use of a particular material may appear labour-intensive at one stage, but impose different requirements at subsequent stages.

These points are illustrated by a comparison of the characteristics of different materials' technological requirements, shown in Table 10.19. The table sheds some light on the question[20] of the association between the labour-intensity of a technique and the product characteristics, i.e. whether low-income products are more labour-intensive than high-income products. Common sense suggests that there *need* be no such association, but the table suggests that there is in fact quite a close association in this particular case: the higher-income products are associated with more investment-intensive techniques, with the exception of stone production, which is a high-income product produced in a labour-intensive way.

The association shown accords with the view that the relationship between techniques and products depends on the historical circumstances of the development of the technology. Traditional methods

TABLE 10.19 Choice of building materials in relation to quality of building, consumer income, source of basic material and labour intensity

| *Material* | *Quality of building* | *Income class*[1] *of consumer* | *Source of basic material*[2] *Place* | *Economic sector* | *Labour intensity*[3] *Material production* | *Processing* | *Building* |
|---|---|---|---|---|---|---|---|
| Mud and wattle | low | subsistence | local | informal | high | high | high |
| Sun-dried clay blocks | low to medium | slightly above subsistence (rural) | local | informal | high | high | high |
| Murram-enforced blocks | low to medium | low (urban) | local | informal formal[4] | high | high-medium | high |
| Black cotton bricks | low to medium | low (urban) | local | informal | high | high | high |
| Stones | high | mainly high | local | informal formal | high | high | high |
| Timber | medium | medium and high | local | informal formal | .[5] | .[5] | .[6] |
| Pre-cast concrete panels | medium | medium and high | local | formal | medium[7] | medium[7] | medium |
| Pre-cast concrete panels made with a foaming agent | medium | medium and high | local and imported | formal | medium[7] | medium[7] | low–medium |
| Cement blocks made with a chemical additive | high | medium and high | local and imported | formal | low | low | low–medium |
| Cement blocks | medium | medium and high | local | formal | medium[7] | medium[7] | medium |

[1] Low: rural subsistence and most small-scale market production, urban unemployed, under-employed – including most informal-sector activities. Medium: unskilled and semi-skilled formal-sector (mostly urban) workers. High: skilled and professional. The use of this classification means that the majority of the population should be classified as low-income.

[2] This refers both to geographical location (local or imported) and to sector of the local economy, which is regarded as being divided into an informal and a formal sector. Broadly, the informal sector is more labour-using, while formal-sector production involves more equipment, often imported.

[3] To classify production methods according to labour intensity implies that each material may be identified with a single technique; as the survey of cement block making showed, that is not the case, and considerable variation may be possible.

[4] Cement.

[5] Intensive use of skilled labour and natural resources.

[6] Intensive use of skilled labour and some use of machinery.

[7] Varies (see footnote 3).

(mud and wattle, and sun-dried blocks) were developed in societies where labour and materials were the only resources available – so naturally they used only these resources; and they were developed for low-income consumers. In contrast, such techniques as pre-cast units with foaming additives, recently developed in advanced countries for consumption there, tend to be relatively investment-intensive and with high-quality characteristics for high-income consumers.

Techniques such as concrete block manufacture, which originated some time ago in developed countries, tend to be intermediate (by modern standards) in terms both of product characteristics and of technology used, reflecting the intermediate factor availability and income levels of advanced countries some decades ago, when the techniques were developed. But of course, there is not a necessary association. Thus craft production is often suitable for high-income consumers, while modern technology can be harnessed to the production of low-income goods.

This brief look at different building materials has thus illustrated some of the complex relationships between choice of product and choice of technique and between income levels and income distribution. For high-quality construction meeting international specifications and demanded by high-income consumers, there are broadly two possibilities, namely the craft-intensive high-quality construction in timber and stone or the mechanised methods of manufacturing pre-cast concrete panels or blocks. Low-income consumers, on the other hand, provide the market for hand-operated cement block manufacture and the various other blocks and bricks formed with primitive technology from local materials. Thus choice of product and technique cannot be divorced from income levels among consumers. Choice of technique also influences income distribution: the mechanised large-scale techniques tend to be associated with relatively high wages, and with profits going to medium-scale to large-scale entrepreneurs. In contrast, the low-income techniques are often used outside the formal sector, and generate low incomes among those who use them.

NOTES

1. For this and other information, including a comprehensive list of quarry operators, I am extremely grateful to E. J. Wells who, in part in cooperation with E. Rado, was conducting a major study of building materials, with particular reference to quarries, in Kenya, at the Institute of Development Studies, from 1966 to 1969. The survey included all the quarry operators who produced blocks in and around Nairobi. Every building firm listed in the yellow pages of the Nairobi telephone directory was contacted and if they owned a block-making machine they were asked for, and in most cases granted, a personal interview. The other organisations interviewed within Nairobi were contacted through the advice of some of the other firms questioned. The organisations outside Nairobi were selected in a much more arbitrary way. I was informed by others (in particular by E. J. Wells and John Anderson) working in connected

fields that various firms operated block-making machines, outside Nairobi and contacted them accordingly.

2. See the inquiry into maize grinding in Chapter 9.
3. One firm (using a stationary vibrating machine) suggested that faulty blocks occurred at the rate of 6/250 for solid blocks and 20/250 for hollow.
4. The quality (uniformity and strength) of the blocks is also affected by the method of mixing and how the mixture is transferred from mixer to machines. In principle any method of mixing and transfer is compatible with any type of block-maker. In practice, because of scale factors, this is not so (see later discussions).
5. Estimates of material costs varied according to location and nature of producer (because material producers tended to price their products at cost price). Cement in Nairobi was said to be around 245/- a ton, compared with 15–20/- a ton for sand and a similar price, 15–20/-, for ballast. The figures given for Nyeri, about 90 miles from Nairobi, were 400/- a ton for cement, 25/- a ton for sand and 10/- for stones.
6. Fuel costs are discussed later. A fairly typical cost was 10 cents a block.
7. In those cases where the owner of the machine repaired it himself he normally did not include his labour as a cost. To the extent that the opportunity cost of his labour was zero he was right, but the skilled labour involved is none the less a real cost of operating such a block-maker.
8. See Chapter 9, note 18 for conversion formula.
9. The term 'broadly' is used because the investment-labour ratios (adjusted) of the laying machine (and of the local adapted machine at low interest rates) are greater than that of the large stationary machine, although the scale of output is smaller in the former categories.
10. Estimates produced by M. Scott (1973) and N. Stern (1972) in their cost/benefit analyses.
11. All these calculations assume a 10 per cent discount rate.
12. The choice of technique in the production of material used has been established elsewhere. For different methods in quarrying see E. J. Wells (1970); for cement, see in particular Doyle (1965) and Diaz-Alejandro (1971).
13. See Jomo Kenyatta (1961), pp. 78–82, for a description of the way in which these houses are constructed by the Kikuyu.
14. A local red earth.
15. In all the methods which use only local labour and no machinery bought from abroad, or from the industrial sector of Kenya, some investment occurs in the sense of delay between the input of the labour and the final output. However, in so far as the scarce investment resources consist of imported equipment or equipment from the industrial sector, this investment does not involve the costs associated with investment-intensive techniques requiring purchased equipment.
16. *Timber housing pilot project at Kariobangi, Nairobi*, report by R. S. Ryatt, Resident Engineer (mimeographed, 1972).
17. Figures from D. Turin: *Notes on housing and construction*, paper prepared for the ILO-organised comprehensive employment strategy mission to Kenya, 1972, para. 5.5.
18. Ibid., para. 5.8.
19. This is also the conclusion to be drawn from W. P. Strassmann's figures, which show a fall in the labour share in construction costs, a rise in materials costs and a rise in absolute costs per square metre, which can of course be offset by reduced land area and costs, as the weight of buildings rises and a switch is made to pre-cast panels. See W. P. Strassmann, (1974).
20. Discussed in Chapter 3.

# 11 Some Conclusions

The concern of this book has been to analyse the role of technology in the economic development of poor countries. What were the effects of the technology adopted? Were these effects inevitable given the nature of the technology available, or could a different choice from within the available range have produced significantly different results? What conditions would be necessary to secure a different choice? What determines the nature of the available technology and how could this be changed? Underlying these questions is a fundamental question about the nature of technology and of technological choice. Rather than provide a detailed step-by-step summary of the book, this chapter will try to focus directly on these questions, in the light of the earlier discussion.

In many poor countries high rates of investment have been associated with slow growth of employment in the modern sector and with maldistribution of income, with much of the fruits of the investment going to the minority associated with the modern sector. A dualistic form of development has emerged, associated with growing un- and under-employment, and concentration of resources on the relatively small sector receiving advanced technology from overseas to the neglect and impoverishment of the rest of the economy. Clearly, therefore, the nature of the investment, in those countries to which this paradigm applies, has been unsatisfactory. The critical question is whether this arises from the nature of the technology itself, as the technological determinists would suggest, or from a wrong choice being made from within the available range.

A fundamental distinction between rich and poor countries is that poor countries are for the most part *recipients* of technology developed in rich countries, while rich countries, as a block, generate their own technology. To examine the nature of the technology available to poor countries therefore involves examining the nature of technological development in rich countries.

Examination of the process of technological development suggests that the range of technology available is narrower than many economists have assumed, but wider than the technological determinist view supposes. The characteristics of a technology are conditioned by the environment in the economy for which it is developed. Given the dominance of advanced countries in technological development, the emerging technology has been in line with the conditions in the advanced countries; its characteristics have been unsuited to the different conditions in the much poorer countries of the third world. Much of the dualistic pattern of development in third world countries can be attributed to the fact that modern technology has characteristics suited to advanced countries, not to poor countries. But capital-intensive techniques from the advanced countries producing sophisticated products do not represent the only alternative. Older technology from advanced countries, traditional technology from third world countries and recently developed technology designed for the third world offer a choice of techniques with different and more appropriate characteristics. But because so little systematic attention has been devoted to appropriate techniques, there are many areas where such techniques are not available and recent technology from advanced countries offers the only alternative. There are other areas where big advances in productivity in technology in advanced countries have made alternative older or traditional techniques obsolete. Empirical case studies show that in many cases there is none the less a significant range of efficient techniques, not just one advanced-country technique with inappropriate characteristics, although often the alternative techniques have relatively low productivity. There has been a strong tendency for empirical studies to be directed towards investigating the range of methods available to produce a given product. Once the investigation is extended to the range of different ways (including different products) of meeting broadly defined needs, the range of available techniques is considerably extended.

If we define the question of choice of technique to include all the different ways of meeting basic needs, then it does appear that there are alternative technologies to those currently in use in advanced countries. From this point of view the technological determinist case is wrong, and poor countries could choose more appropriate technology. But the dominating position of R and D in rich countries has meant that the productivity of the alternative technology is often relatively low, while the range of products it is suited to is somewhat limited. Given then that poor countries do have some choice – and the extent of the choice they have is still being explored by empirical studies – it follows that the determinants of that choice, here described as the selection mechanisms, are partly responsible for the nature of the technology adopted.

The question then arises of what determines the choice of technique – the question which has been central to much analysis of technology. Neo-classicists have always emphasised relative prices as being of critical significance. But the analysis of technical choice made here suggests a much more complicated process, with relative prices being just one of the selection mechanisms, often of only minor significance.

Each technique has a vector of characteristics consisting of such things as product type, quality, scale of output, resource use and so on. Decision makers differ in objectives and in constraints, including knowledge. The actual choice in a particular case then depends on the interaction between decision makers' objectives, given the constraints they face and the characteristics of different techniques. At a macro level, the choice of techniques depends on the weight of investment accounted for by different types of decision makers, as well as the decisions they each make.

From this type of approach to technical choice a set of significant selection mechanisms emerges. The set includes the technology already in use. Technology is a package, and each technique is designed to be operated within a particular technological system. The adoption of one technique from an alternative system is in many cases only efficient if the associated parts of the technological system are also adopted – either produced locally, or imported. The technology in use therefore helps determine the choice of technique in associated processes.

The nature of the markets served and the type of product required are of critical significance in determining the nature of the technology. International markets, particularly markets in the advanced countries, and demand from the upper income groups in poor countries who have adopted advanced-country consumption patterns, require for the most part recent products designed in advanced countries, and therefore recent technology. International trading policies, and local income distribution, as well as policies towards advertising and product standards are important influences on choice of technique. Relative prices and availability of resources – including knowledge – also play a role.

Different types of decision maker – multinational companies, local public and local private companies, large or small enterprises, organised on Western or traditional lines – tend to make different choices. This is because they have different objectives and different opportunities and constraints. Their access to different types of technology differs, as do the markets they face and their access to other resources such as finance, skilled and unskilled labour. Thus for the economy as a whole the proportion of total resources controlled by different types of decision maker heavily influences the choice of technique.

The extent to which governments may *choose* to pursue alternative paths depends on the extent to which they control these (and other) selection mechanisms. Many of the selection mechanisms are, in one way or another, the *outcome* of the economic system, and of the technology already in use. For example, income distribution and the weight of control over resources of different types of decision makers in part, at least, emerge from the economic system. Thus a system based on the use of advanced-country technology tends to generate high incomes among those employed with the technology, providing markets for the goods the technology produces. Again, on the technical side, because of the links between different techniques, the technology in use partly determines the relative advantage of different types of technique in new investment decisions. Relative prices and availability of resources are also in part the outcomes of the existing system. Advanced-country technology is strongly associated with relatively high wages whereas traditional technology is associated with low incomes. Hence, given the nature of the existing technology, governments may have only limited control over relative prices.

In all these areas the causation runs both ways, with a particular technological system giving rise to selection mechanisms which are consistent with it, leading to new decisions of a similar kind. While there are links of this kind which occur in an unregulated situation, the causal links may be loose, so that government intervention could push the selection mechanisms in different directions. How far can governments break through the links so as to achieve this?

This question needs to be considered at two levels. One is the level of political economy. Governments are normally the representatives of a particular political economy, rather than its arbiters. To the extent that governments consist of individuals who benefit from, and represent those who benefit from, the political economy in being, they may not wish or be able to challenge it. An alternative technology at a macro level involves an alternative political economy – a different distribution of the benefits of the economic system. Governments which have developed in one system may not be powerful enough to choose an alternative system. This does not mean that they will not promote alternative techniques in a few cases where the benefits are obvious and the attack on existing interests is insignificant. But it does impose limits on their willingness and ability to pursue different policies on a sufficient scale to secure a significant change at the macro level.

The effective pursuit of an alternative appropriate technology would threaten interests in the advanced countries as well as those in the underdeveloped countries who are currently benefiting from the use of advanced-country technology. The continued use of advanced-country technology is at the heart of the continued dependence of the poor countries. It maintains the advanced countries' lead in technology and therefore permits, indeed necessitates, the continued sale of technology, goods and managerial services to poor countries, on terms favourable to the rich countries. There thus develops a sort of alliance of interests between those in the third world who benefit from the continued use of advanced-country technology – those who own or work with the technology and receive high incomes from its high productivity – and those in the advanced countries who gain by the maintenance of technological dependence in poor countries. Dependency theorists have formalised this relationship, believing that governments in dependent countries represent the interests of the centre in the periphery and for that reason are incapable of challenging it. This is perhaps a too simplistic view – some governments in poor countries do take action which appears to be counter to interests in advanced countries. Moreover, some governments show interest in the welfare of groups outside the privileged sectors. None the less, the alliance of interests between advanced countries and the advanced technology sector in poor countries presents a formidable obstacle to any significant change.

The second level is the technical level. Given political will, technical problems may prevent an effective change in selection mechanisms towards an alternative technology. As already suggested, many of the selection mechanisms are the outcome of the economic/technological system in being and not in the control of governments. This is clearly true, for example, of the technical requirements on complementary processes imposed by the existing technology. Moreover, many aspects of the system, while not strictly essential to it, are conducive to the system's efficiency. This is true of much of the concentration of resources in the advanced sector that tends to be associated with the use of advanced-country technology: the concentration of expenditure on the social services and the infrastructure is required for the efficient working of the system; the relatively high wages paid by the sector provide for an efficient workforce and also markets for the products the technology produces. To change these aspects, as is required for an alternative technology, would not only meet political opposition but would also threaten the efficiency of the existing system.

There are some conditions which are likely to make a move to an alternative path easier. One is the international trading system. It was argued (Chapter 8) that a move away from trade with advanced countries towards trade with other developing countries is a necessary condition for the pursuit of a more appropriate technology. The development of local capital goods capacity is another, allowing the production of more appropriate machines. What is needed above all is local technical innovation directed towards local needs. The increasing relative efficiency of advanced-country technology due to the dominating size of resources devoted to it makes the choice of alternative techniques increasingly costly. To secure technical innovation appropriate to the third world requires much more than a formal structure of local scientific institutions, as has been powerfully argued elsewhere.[1] It also requires that decision makers *use* this sort of technology, which means that the selection mechanisms have to be right. But this requires economic changes which are only likely to occur *after* an alternative technology has been successfully adopted, and are unlikely to precede it.

We are coming close to a different type of technological determinism. According to the old-style technological determinism, advanced-country technology is used in poor countries because it represents the only efficient technology. This position has been rejected on two grounds: first, because alternative techniques do exist, particularly if we include all the different ways of meeting broadly defined needs. Secondly, because technical change is itself the product of an economic system and may therefore be directed to produce an alternative technology. Thus a potential choice of technology does exist both in relation to current techniques and to the direction of technical change. But the way in which the choice is made is determined by an economic system which itself depends on the technology in use. The situation is technologically determined not because there is no alternative technology to choose from but because of the complex links between the selection mechanisms and the techniques chosen, which make the choice of an alternative system extremely difficult.

NOTES

1. See Cooper (1972), and (1974).

# Bibliographical References


Aaronovitch, S., and M. C. Sawyer, 'The Concentration of British Manufacturing', *Lloyds Bank Review*, no. 14 (October 1974).

Adelman, I., and C. T. Morris, *Economic Growth and Social Equity in Developing Countries* (Stanford University Press, 1973).

Amin, S., *Neo-Colonialism in West Africa* (Penguin, 1973).

Armstrong, A., and A. Silberston, 'Size of Plant, Size of Enterprise and Concentration in British Manufacturing Industry, 1935–1958', *Journal of the Royal Statistical Society*, vol. 128, series A (1965).

Arrighi, G., and J. S. Saul, *Essays on the Political Economy of Africa* (Monthly Review Press, 1973).

Arrow, K. J., H. B. Chenery, B. S. Minhas and R. M. Solow, 'Capital-Labor Substitution and Economic Efficiency', *Review of Economics and Statistics*, vol. XLIII (August 1961).

Arrow, K. J., 'The Economic Implications of Learning by Doing', *Review of Economic Studies*, vol. XXIX (June 1962).

Atkinson, A. B., and J. B. Stiglitz, 'A New View of Technological Change', *Economic Journal*, vol. LXXIX (September 1969).

Baer, W., and M. Hervé, 'Employment and Industrialisation in Developing Countries', *Quarterly Journal of Economics*, vol. LXXX (February 1966).

Bain, J. S., *Barriers to New Competition* (Harvard University Press, 1956).

Baran, P., *The Political Economy of Growth* (Monthly Review Press, 1957).

Barna, T., 'The Replacement Cost of Fixed Assets in British Manufacturing Industry in 1955', *Journal of the Royal Statistical Society*, vol. 120, series A (1957).

Baron, C. G., 'Sugar Processing Techniques in India', in Bhalla (ed.) 1975.

Baumol, W. J., *Economic Theory and Operations Analysis* (Prentice-Hall, 1961).

Bennett, R. L., 'Surplus Agricultural Labor and Development – Facts and Theories: Comment', *American Economic Review*, vol. LVII (March 1967).

Berger, P. L., B. Berger and H. Kellner, *The Homeless Mind* (Random House, 1973; Pelican, 1974).

Bernal, J., *Science in History* (Penguin, 1969).

Bhagwati, J. N., 'The Pure Theory of International Trade: A Survey', *Economic Journal*, vol. LXXIV (March 1964).

Bhagwati, J. N., and S. Chakravarty, 'Contributions to Indian Economic Analysis: A Survey', Supplement to *American Economic Review*, vol. LIX, part 2 (September 1969).

Bhagwati, J. N., and P. Desai, *India, Planning for Industrialisation* (Oxford University Press, 1970).

Bhalla, A. S., 'Galenson-Leibenstein Criterion of Growth Reconsidered: Some Implicit Assumptions', *Economia Internazionale*, vol. XVII (1964(a)).

Bhalla, A. S., 'Investment Allocation and Technological Choice – A Case of Cotton Spinning Techniques', *Economic Journal*, vol. LXXIV (September 1964(b)).

Bhalla, A. S., 'Choosing Techniques: Hand Pounding v. Machine Milling of Rice: An Indian Case', *Oxford Economic Papers*, vol. 17 (March 1965).

Bhalla, A. S., 'Techniques in the Construction Industry in China and India', *World Development*, vol. 2 (March 1974).

Bhalla, A. S. (ed.), *Technology and Employment in Industry* (I.L.O., 1975).

Bhalla, A. S., and J. Gaude, 'Appropriate Technologies in Services with Special Reference to Retailing' (I.L.O., mimeo, 1973).

Bhalla, A. S., and F. Stewart, 'International Action for Appropriate Technology', Background Paper for the World Employment Conference (Geneva: I.L.O., 1976).

Blaug, M., P. Layard, and M. Woodhall, *The Causes of Graduate Unemployment in India* (Allen Lane, 1969).

Bloom, G. F., 'Union Wage Pressure and Technological Discovery', *American Economic Review*, vol. XLI (September 1951).

Boon, G. K., *Economic Choice of Human and Physical Factors in Production* (North Holland Publishing Co., 1964).

Boon, G. K., 'Technological Choice in Metalworking with Special Reference to Mexico', in Bhalla (ed.) 1975.

Boserup, E., *Conditions of Agricultural Growth* (Allen & Unwin, 1965).

Boserup, E., *Woman's Role in Economic Development* (Allen & Unwin, 1970).

Bruno, M., 'Development Policy and Dynamic Comparative Advantage', in R. Vernon (ed.), *The Technology Factor in International Trade* (National Bureau of Economic Research, 1970).

Bruton, H., 'Economic Development and Labour Use: A Review', *World Development*, vol. 1 (December 1973); and in Edwards (ed.) 1974.

Carnoy, M., and H. H. Thias, 'Rates of Return to Schooling in Kenya', *Eastern Africa Economic Review*, vol. 3 (December 1971).

Carpenter, Frank W. (Chairman), *Report of the Committee on African Wages* (Nairobi: Government Printer, 1954).

Carr, M., *Economically Appropriate Technologies for Developing Countries: an Annotated Bibliography* (Intermediate Technology Development Group, 1976).

Chandavarkar, A. G., 'Interest Rate Policies in Developing Countries', *Finance and Development*, vol. 7 (March 1970).

Chenery, H. B., 'Process and Production Functions for Engineering Data', in W. Leontief (ed.), *Studies in the Structure of the American Economy* (Oxford University Press, 1953).

Chenery, H., M. S. Ahluwalia, C. L. G. Bell, J. H. Duloy and R. Jolly, *Redistribution with Growth* (Oxford University Press, 1974).

Chenery, H. B. and M. Bruno, 'Development Alternatives in an Open Economy: The Case of Israel', *Economic Journal*, vol. LXXII (March 1962).

Cooper, C., 'Science Policy and Technological Change in Underdeveloped Economies', *World Development*, vol. 2 (March 1974).

Cooper, C., and R. Kaplinsky, *Second-Hand Equipment in a Developing Country* (I.L.O. 1974).

Cooper, C., R. Kaplinsky, R. Bell and W. Satyarakwit, 'Choice of Techniques for Can Making in Kenya, Tanzania and Thailand', in Bhalla (ed.) 1975.

Cooper, C., and F. Sercovich, 'The Channels and Mechanisms for the Transfer of Technology from Developed to Developing Countries' (UNCTAD paper TD/D/AC. 11/5, 1970).

Court, D., and D. P. Ghai, (eds), *Education, Society and Development: New Perspectives from Kenya* (Oxford University Press, 1974).

Das Gupta, P., S. Marglin and A. K. Sen, *Guidelines for Project Evaluation* (UNIDO, 1972).

Dean, A., 'The Stock of Fixed Capital in the United Kingdom', *Journal of the Royal Statistical Society*, vol. 127, series A (1964).

Dell, S., *Trade Blocs and Common Markets* (Constable, 1963).

Denison, E., *Why Growth Rates Differ* (Brookings Institution, 1967).

Desai, M., and D. Mazumdar, 'A Test of the Hypothesis of Disguised Unemployment', *Economica*, vol. XXXVII (February 1970).

Dhar, P. N., and H. F. Lydall, *The Role of Small Enterprises in Indian Economic Growth* (Asia Publishing House, 1961).

Diaz-Alejandro, C. F., 'Industrialisation and Labour Productivity Differentials' *Review of Economics and Statistics*, vol. XLVII (May 1965).

Dickson, D., *Alternative Technology and the Politics of Technical Change* (Fontana, 1974).

Dobb, M., 'Second Thoughts on Capital Intensity of Investment', *Review of Economic Studies*, vol. XXIV (1956–7).

Doctor, K. C., and H. Gallis, 'Modern Sector Employment in Asian

Countries: Some Empirical Estimates', *International Labour Review*, vol. XC, No. 6 (December 1964).

Doctor, K. C., and H. Gallis, 'Size and Characteristics of Wage Employment in Africa: Some Statistical Estimates', *International Labour Review*, vol. 93 (February 1966).

Domar, E., *Essays in the Theory of Growth* (Oxford University Press, 1957).

Dovring, F., 'The Share of Agriculture in a Growing Population', *Monthly Bulletin of Agricultural Economics and Statistics* (F.A.O., August/September 1959).

Doyle, L. A., *Inter-Economy Comparisons – A Case Study* (University of California Press, 1965).

Eckaus, R. S., 'The Factor Proportions Problem in Underdeveloped Areas', *American Economic Review*, vol. XLV (September 1955).

Economic Commission for Latin America (ECLA), *Choice of Technique in the Latin American Textile Industry* (U.N., 1966).

Edwards, E. O. (ed.), *Employment in Developing Countries* (Colombia University Press, 1974).

Elkan, W., *Migrants and Proletarians: Urban Labour in the Economic Development of Uganda* (Oxford University Press, 1960).

El-Karanshawy, H. A. S., *Choice of Carpet Weaving Technology in Egypt* (Ph. D. thesis, University of Strathclyde, 1975).

Emmanuel, A., *Unequal Exchange: A Study of the Imperialism of Trade* (Monthly Review Press, 1972).

Enos, J. L., 'More (or Less) on the Choice of Technique, with a Contemporary Example', (Magdalon, Oxford, mimeo, 1974).

Erlich, A., *The Soviet Industrialisation Debate 1924–1928* (Harvard University Press, 1960).

Fei, J. C. H., and G. Ranis, 'A Model of Growth and Employment in the Open Dualistic Economy: the Cases of Korea and Taiwan' in Stewart (ed.) 1975.

Fields, G. S., 'Private Returns and Social Equity in the Financing of Higher Education', in Court and Ghai (eds) 1974.

Frank, A. G., *Capitalism and Underdevelopment in Latin America* (Monthly Review Press, 1967).

Frank, A. G., *Latin America, Underdevelopment or Revolution* (Monthly Review Press, 1969).

Freeman, C., 'Research and Development: a Comparison between British and American Industry', *National Institute Economic and Social Review*, no. 20 (May 1962).

Freeman, C., and associates, 'Chemical Process Plant, Innovation and the World Market', *National Institute Economic Review*, no. 45 (August 1968).

Furtado, C., *Development and Underdevelopment* (University of California Press, 1964).

Galbraith, J. K., *The New Industrial State* (Hamish Hamilton, 1967).

Galenson, W., and H. Leibenstein, 'Investment Criteria, Productivity and Economic Development', *Quarterly Journal of Economics*, vol. LXIX (August 1955).

Ganesan, S., 'Employment Generation through Investments in Housing and Construction', (Ph.D. thesis, London University, 1975).

Gerschenkron, A., *Economic Backwardness in Historical Perspective* (Harvard University Press, 1962).

Ghai, D. P., 'Incomes Policy in Kenya: Need, Criteria and Machinery', *East African Economic Review*, vol. 4, new series (December 1968).

Ghai, D. P. (ed.), *Economic Independence in Africa* (East African Literature Bureau, 1973).

Ghai, D. P., 'Towards a National System of Education', in Court and Ghai (eds) 1974.

Gouverneur, J., *Productivity and Factor Proportions in Less Developed Countries: The Case of Industrial Firms in the Congo* (Oxford University Press, 1971).

Granick, D., *Soviet Metal Fabricating and Economic Development* (University of Wisconsin Press, 1967).

Griffin, K., *Underdevelopment in Spanish America* (Allen & Unwin, 1969).

Griffin, K., 'The International Transmission of Inequality', *World Development*, vol. 2 (March 1974).

Gustafson, W. E., 'Research and Development, New Products and Productivity Change', *American Economic Review*, Papers and Proceedings, vol. LII (1962).

Habakkuk, H. J., *American and British Technology in the Nineteenth Century*, (Cambridge University Press, 1962).

Harcourt, G. C., *Some Cambridge Controversies in the Theory of Capital* (Cambridge University Press, 1972).

Harris, J. R., and M. P. Todaro, 'Wages, Industrial Employment and Labour Productivity: the Kenyan Experience', *Eastern Africa Economic Review*, vol. I (June 1969).

Harris, J. R., and M. P. Todaro, 'Migration, Unemployment and Development', *American Economic Review*, vol. LX (March 1970).

Hart, K., 'Informal Income Opportunities and Urban Employment in Ghana', paper delivered to Conference on Urban Unemployment in Africa, Institute of Development Studies, Sussex University, September 1971, partially printed in Jolly *et al.* (1973).

Hayek, F. A., and R. Lekachman (eds), *National Policy for Economic Welfare at Home and Abroad* (Doubleday, 1955).

Healey, J. M., 'Industrialisation, Capital Intensity and Efficiency', *Bulletin of the Oxford Institute of Economics and Statistics*, vol. 30 (November 1968).

Heckscher, E. F., 'The Effect of Foreign Trade on the Distribution of

Income', reprinted in H. S. Ellis and L. A. Metzler (eds), *Readings in the Theory of International Trade* (American Economic Association, 1949).

Helleiner, G. K., 'Manufactured Exports from Less Developed Countries and Multinational Firms', *Economic Journal*, vol. 83 (March 1973).

Helleiner, G. K., 'The Role of Multinational Corporations in the Less Developed Countries' Trade in Technology', *World Development*, vol. 3 (April 1975).

Her Majesty's Stationery Office, *National Accounts – Sources and Methods*, no. 13 (1968).

Hinchcliffe, K., 'Education, Individual Earnings and Earnings Distribution' in Stewart (ed.), 1975.

Hirsch, S., *Location of Industry and International Competitiveness* (Clarendon Press, 1967).

Hirsch, S., 'Trade and Per Capita Income Differentials: A Test of the Burenstam-Linder Hypothesis', *World Development*, vol. 1 (September 1973(a)).

Hirsch, S., 'Hypotheses regarding Trade between Developing and Industrialised Countries', mimeo (June 1973(b)).

Hirschman, A., *The Strategy of Economic Development* (Yale University Press, 1958).

Hopkins, A. G., *An Economic History of West Africa* (Longmans, 1973).

Hufbauer, G. C., 'The Impact of National Characteristics and Technology on the Commodity Composition of Trade in Manufactured Goods', in R. Vernon (ed.), *The Technology Factor in International Trade* (National Bureau of Economic Research, 1970).

Hufbauer, G. C., and J. G. Chilas, 'Specialisation by Industrial Countries: Extent and Consequences', Conference on Problems of International Division of Labour (Institut für Weltwirtschaft an der Universität Kiel, 1973).

Hutton, C., 'Aspects of Urban Unemployment in Uganda,' East African Institute of Social Research Conference Papers (January 1966).

Huxley, E., *Flame Trees of Thika* (Chatto & Windus, 1959).

International Labour Office, *Employment, Incomes and Equality: A Strategy for Increasing Productive Employment in Kenya* (Geneva, 1972).

International Labour Office, *Measurements of Underemployment, Concepts and Methods*, Report IV (Geneva, 1966).

International Labour Office, *Report to the Government of the United Republic of Tanzania on Wages, Incomes and Prices Policy* (Dar-es-Salaam: Government Printer, 1967).

International Labour Office, *Towards Full Employment* (Geneva, 1970).

Islam, N., 'Comparative Costs, Factor Proportions and Industrial Efficiency in Pakistan', *Pakistan Development Review*, vol. VII summer, 1967).

James, J., 'Products, Processes and Incomes: Cotton Clothing in India', *World Development*, vol. 4 (February, 1976).

Jenkins, G., *Non-Agricultural Choice of Technique, An Annotated Bibliography of Empirical Studies* (Oxford: Institute of Commonwealth Studies, 1975).

Johnson, H. G., *International Trade and Economic Growth* (Allen & Unwin, 1958).

Johnson, H. G., *Comparative Cost and Commercial Policy, Theory for a Developing World Economy*, Wicksell Lectures (1968).

Johnson, W. A., *The Steel Industry of India* (Harvard University Press, 1966).

Johnston, B. F., *Agriculture and Economic Development: The Relevance of the Japanese Experience*, Food Research Institute Studies, Stanford University, vol. VI (1966).

Jolly, R., E. de Kadt, H. Singer and F. Wilson, *Third World Employment* (Penguin, 1973).

Joshi, H., 'World Prices as Shadow Prices', *Bulletin of the Oxford University Institute of Economics and Statistics*, vol. 34 (February 1972).

Joshi, V., 'Saving and Foreign Exchange Constraints', in P. Streeten (ed.), *Essays in Honour of Lord Balogh* (Weidenfeld & Nicolson, 1970).

Kaldor, N., 'A Model of Economic Growth', *Economic Journal*, vol. LXVII (December 1957).

Kaldor, N., in R. Robinson (ed.), *Industrialisation in Developing Countries* (Cambridge Overseas Studies Committee, 1965).

Kao, C. H. D., K. Anschel and C. K. Eicher, 'Disguised Unemployment in Agriculture: A Survey', in C. K. Eicher and L. Witt (eds), *Agriculture in Economic Development*, (McGraw Hill, 1964).

Kaplinsky, R., 'Innovation in Gari Production: The Case for an Intermediate Technology', I.D.S. Discussion Paper (Sussex University, 1974).

Kay, C., 'Comparative Development of the European Manorial System and the Latin American Hacienda System', *Journal of Peasant Studies*, vol. 2 (October 1974).

Khan, A. R., and A. MacEwan, 'A Multisectoral Analysis of Capital Requirements for Development Planning in Pakistan', *Pakistan Development Review*, vol. VII (winter 1967).

Kidron, M., *Capitalism and Theory* (Pluto Press, 1974).

Kilby, P., *African Enterprise: The Nigerian Bread Industry* (Stanford University Press, 1965).

Kilby, P., 'Industrial Relations and Wage Determination: Failure of the Anglo-Saxon Model', *Journal of Developing Areas*, vol. 1 (July 1967).

King, K., 'Kenya's Informal Machine Makers', *World Development*, vol. 2 (April/May 1974).

Kinyanjui, K., 'Education, Training and Employment of Secondary School Leavers in Kenya', in Court and Ghai (eds) 1974.

Koji, T., 'Wage Differentials in Developing Countries: A Survey of Findings', *International Labour Review*, vol. 93 (March 1966).

Krishnamurty, J., 'Some Aspects of Unemployment in Urban India', in Stewart (ed.) 1975.

Kuznets, S., 'The Gap: Concepts, Measurements, Trends', in G. Ranis (ed.), *The Gap Between Rich and Poor Nations* (Macmillan, 1972).

Lal, D., 'Foreign Exchange Constraints in Economic Development', *Indian Economic Journal* (July/September 1970).

Lal, D., *Men or Machines, A Philippine Case Study of Labour-Capital Substitution in Road Construction* (I.L.O., Geneva, 1977).

Lall, S., 'Transfer-Pricing by Multinational Manufacturing Firms', *Bulletin of Oxford University Institute of Economics and Statistics*, vol. 35 (August 1973).

Lancaster, K., 'A New Approach to Consumer Theory', *Journal of Political Economy*, vol. LXXIV (April 1966).

Lancaster, K., 'Socially Optimal Product Differentiation', *American Economic Review*, vol. LXV (September 1975).

Landes, D. S., *The Unbound Prometheus* (Cambridge University Press, 1969).

Langdon, S., 'Multinational Corporations' Taste Transfer and Underdevelopment: A Case Study from Kenya', *Review of African Political Economy* (January-April, 1975).

Lary, H. B., *Imports of Manufactures from Less Developed Countries* (National Bureau of Economic Research, 1968).

Leff, N. H., *The Brazilian Capital Goods Industry 1929–1964* (Harvard University Press, 1968).

Lewis, W. A., *The Theory of Economic Growth* (Allen & Unwin, 1955).

Linder, S. B., *An Essay on Trade and Transformation* (Almqvist & Wicksell, 1961).

Lipsey, R. G., *The Theory of Customs Unions: A General Equilibrium Analysis*, (Weidenfeld & Nicolson, 1970).

Little, I. M. D., and J. A. Mirrlees, *Manual of Industrial Project Analysis* (O.E.C.D., 1969).

Little, I. M. D., T. Scitovsky and M. Scott, *Industry and Trade in Some Developing Countries* (O.E.C.D. and Oxford University Press, 1970).

Mahalanobis, P. C., 'Some Observations on the Process of Growth in National Income', *Sankhya*, vol. 12, part 4 (1953).

Maizels, A., *Industrial Growth and World Trade* (Cambridge University Press, 1963).

Marglin, S. A., 'What do Bosses do? The Origins and Functions of Hierarchy in Capitalist Production', *Review of Radical Political Economics*, vol. VI (summer 1974).

Maritano, N., *A Latin American Common Market* (University of Notre Dame Press, 1970).

Marris, P., and A. Somerset, *African Businessmen, A Study of Entrepreneurship in Kenya* (Routledge & Kegan Paul, 1971).

Marsden, K., 'Progressive Technologies for Developing Countries', in W. Galenson (ed.) *Essays on Employment* (I.L.O., 1971).

Marx, K., *Capital*, Lawrence & Wishart edition (1970).

Mason, R. Hal, *The Transfer of Technology and the Factor Proportions Problem. The Philippines and Mexico*, UNITAR Research Project no. 10 (1970).

McClelland, D., and D. Winter, *Motivating Economic Achievement* (Free Press, 1969).

McKinnon, R. I., *Money and Capital in Economic Development* (Brookings Institution, 1973).

McRobie, G., and M. Carr, 'Mass Production or Production by the Masses? Technology: A Critical Choice for Developing Countries' (Intermediate Technology Development Group, 1975).

Meade, J. E., *A Neo-Classical Theory of Economic Growth* (Unwin University Books, 1961).

Meier, G., *The International Economics of Development, Theory and Policy* (Harper & Row, 1968).

Merhav, M., *Technological Dependence, Monopoly and Growth* (Pergamon Press, 1969).

Merrett, A. J., and A. Sykes, *The Finance and Analysis of Capital Projects* (Longmans, 1963).

Merrill, R. S., contribution in D. L. Sills (ed.), *Encyclopaedia in Social Sciences* (Macmillan, 1968).

Morawetz, D., 'Employment Implications of Industrialisation in Developing Countries: A Survey', *Economic Journal*, vol. 84 (September 1974).

Morawetz, D., 'Elasticities of Substitution in Industry,' *World Development*, vol. 4, no. 1 (January 1976).

Morley, S. A., and G. W. Smith, 'Managerial Discretion and the Choice of Technology by Multinational Firms in Brazil', Program of Development Studies, Rice University, Paper no. 56 (fall 1974).

Mureithi, L. P., 'Demographic and Technological Variables in Kenya's Employment Scene', *Eastern Africa Economic Review*, vol. 6 (June 1974).

Myrdal, G., *Economic Theory and Underdeveloped Regions* (Duckworth, 1957).

Myrdal, G., *Asian Drama* (Penguin, 1968).

National Council of Applied Economic Research (NCAER), *Study of Selected Small Industrial Units* (New Delhi, 1972).

Nayyar, D., *India's Exports and Export Policies since 1960* (Cambridge University Press, 1976).

Nurske, R., *Problems of Capital Formation in Underdeveloped Countries* (Blackwell, 1953).

O'Herlihy, C. St. J., 'Capital/Labour Substitution and the Developing Countries: A Problem of Measurement', *Bulletin of Oxford University Institute of Economics and Statistics*, vol. 34 (August 1972).

Ohlin, B., *Interregional and International Trade*, rev. ed. (Harvard University Press, 1967).

Okita, S., 'Choice of Techniques: Japan's Experience and its Implications', in K. Berrill (ed.), *Economic Development with Special Reference to East Asia* (Macmillan, 1964).

Ozawa, T., *Imitation, Innovation and Trade: A Study of Foreign Licensing Operations in Japan* (Colombia University Ph.D., 1966).

Ozawa, T., *Transfer of Technology from Japan to Developing Countries*, UNITAR Research Report no. 7 (1971).

Pack, H., 'The Employment-Output Trade-Off in LDCs – A Microeconomic Approach', *Oxford Economic Papers*, vol. 26 (November 1974).

Pack, H., 'The Choice of Technique and Employment in the Textile Industry', in Bhalla (ed.) 1975.

Pack, H., and M. P. Todaro, 'Technical Transfer, Labour Absorption and Economic Development', *Oxford Economic Papers*, vol. 21 (November 1969).

Paglin, M., 'Surplus Agricultural Labour and Development: Facts and Theories', *American Economic Review*, vol. LV (September 1965).

Paglin, M., 'Reply', *American Economic Review*, vol. LVII (March 1967).

Peck, M., 'Innovations in the Post-War Aluminium Industry' in *The Rate and Direction of Inventive Activity* (National Bureau of Economic Research, 1962).

Penrose, E., *The Theory of the Growth of the Firm* (Blackwell, 1959).

Pfefferman, G., *Industrial Labour in the Republic of Senegal* (Praeger, 1968).

Phelps, E. S., 'Substitution, Fixed Proportions, Growth and Distribution', *International Economic Review*, vol. 4 (September 1963).

Phelps Brown, E. H., 'Levels and Movements of Industrial Productivity and Real Wages Internationally Compared 1860–1970', *Economic Journal*, vol. 83 (March 1973).

Pickett, J., D. Forsyth and N. McBain, 'The Choice of Technology, Economic Efficiency and Employment in Developing Countries', *World Development*, vol. 2, (March 1974).

Pickett, J., and R. Robson, 'Technology and Employment in the Production of Cotton Cloth', *World Development* (forthcoming, 1977).

Poeplau, W., and C. Schlage, *Nutrition and Health in Usambara* (1966).

Posner, M. V., 'International Trade and Technical Change', *Oxford Economic Papers*, vol. 13 (October 1961).

Prasad, K., *Technological Choice under Developmental Planning* (Popular Prakashan, 1963).

Pratten, C. F., *Economies of Scale in Manufacturing Industry* (Cambridge University Press, 1971).

Prebisch, R., *Towards a New Trade Policy for Development* (U.N., 1964).

Preobrazhensky, E., *The New Economics* (1926; Oxford University Press translation, 1965).

Rado, E., 'An Explosive Model of Education', *Manpower and Unemployment Research in Africa*, vol. 6 (November 1973).

Raj, K. N., 'Employment and Unemployment in the Indian Economy: Problems of Classification, Measurement and Policy', *Economic Development and Cultural Change*, vol. VII (April 1959).

Raj, K. N., 'Growth Models in Indian Planning', *Indian Economic Review*, vol. 5 (February 1961).

Raj, K. N., and A. K. Sen, 'Alternative Patterns of Growth under Conditions of Stagnant Export Earnings', *Oxford Economic Papers*, vol. 13 (February 1961).

Ranis, G., 'Output and Employment in the 70's: Conflicts or Complements', in R. Ridker and H. Lubell (eds), *Employment and Unemployment Problems of the Near-East and South Asia* (Vikas Publications, 1971).

Ranis, G., 'Technology, Employment and Growth: the Japanese Experience', in *Automation in Developing Countries* (I.L.O., 1972(a)).

Ranis, G., 'Some Observations on the Economic Framework for Optimum LDC Utilisation of Technology', in USAID, *Technology and Economics in International Development, Report of a Seminar* (May 1972(b)).

Ranis, G., 'Industrial Sector Labour Absorption', *Economic Development and Cultural Change*, vol. 21 (April 1973).

Redfern, P., 'Net Investment in Fixed Assets in the United Kingdom, 1938–53', *Journal of the Royal Statistical Society*, vol. 118, series A (1955).

Robana, A., *The Prospects for an Economic Community in North Africa, Managing Economic Integration in the Mahgreb States* (Praeger, 1973).

Robertson, D. H., *Britain in the World Economy* (Allen & Unwin, 1954).

Robinson, J., *Essays in the Theory of Employment* (Macmillan, 1937).

Robinson, J., 'The Production Function and the Theory of Capital', *Review of Economic Studies*, vol. XXI (1953–4).

Robson, P. (ed.), *International Economic Integration* (Penguin Readings, 1971).

Rosenberg, N., 'Capital Goods, Psychology and Economic Growth', *Oxford Economic Papers*, vol. 15 (November 1963(a)).

Rosenberg, N., 'Technological Change in the Machine Tool Industry,

1840–1910', *Journal of Economic History*, vol. XXIII (December 1963(b)).

Rosenberg, N., 'The Direction of Technological Change: Inducement Mechanisms and Focussing Devices', *Economic Development and Cultural Change*, vol. 18, (October 1969).

Rostow, W. W. *The Stages of Economic Growth: A Non-Communist Manifesto* (Cambridge University Press, 1960).

Sabot, R., 'The Meaning and Measurement of Urban Surplus Labour in an African Context' (Oxford University Institute of Economics and Statistics, mimeo, 1974).

Salter, W. E. G., *Productivity and Technical Change*, 2nd ed. (Cambridge University Press, 1966).

Samuelson, P., 'International Trade and the Equalisation of Factor Prices', *Economic Journal*, vol. LVIII (June 1948).

Sandesara, J. C., 'Scale and Technology in Indian Industry', *Bulletin of Oxford University Institute of Economics and Statistics*, vol. 28 (August 1966).

Sands, S., 'Changes in the Scale of Production in U.S. Manufacturing Industry, 1904–1947', *Review of Economics and Statistics*, vol. XLIII (November 1961).

Dos Santos, T., 'The Crisis of Development Theory and the Problem of Dependence in Latin America' (Siglio, 1969; reprinted in H. Bernstein (ed.), *Underdevelopment and Development* (Penguin Readings, 1973)).

Saul, S. B., 'The Market and Development of the Mechanical Engineering Industries in Britain, 1860–1914', *Economic History Review*, vol. XX, 2nd series (1967).

Sawyer, M. C., 'Concentration in British Manufacturing Industry', *Oxford Economic Papers*, vol. 23 (November 1971).

Schlage, C., 'Polished versus Whole Maize: Some Nutritional and Economic Implications of the Traditional Processing of Maize in North Eastern Tanzania', Bureau of Resource Assessment and Land Use Planning, Research Report no. 2, (Dar-es-Salaam, July 1968).

Schmookler, J., *Invention and Economic Growth* (Harvard University Press, 1966).

Schumacher, E. F., 'Not Just Poverty . . . Misery', *Enterprise* (1969).

Schumacher, E. F., *Small is Beautiful: A Study of Economics as if People Mattered*, (Blond & Briggs, 1973).

Scott, M., 'Estimates of Shadow Wages in Kenya' (Nuffield, Oxford, mimeo, 1973).

Sen, A. K., 'Working Capital in the Indian Economy: A Conceptual Framework and Some Estimates', in P. N. Rosenstein-Rodan (ed.), *Pricing and Fiscal Policies*, (Allen & Unwin, 1964).

Sen, A. K., 'Peasants and Dualism with or without Surplus Labour', *Journal of Political Economy*, vol. LXXIV (Oxtober 1966).

Sen, A. K., *Choice of Techniques*, 3rd ed. (Blackwell, 1968).

Sen, A. K., 'On Some Debates in Capital Theory', in A. Mitra (ed.), *Economic Theory and Planning: Essays in Honour of A. K. Das Gupta* (Oxford University Press, 1974).

Sen, A. K., *Employment, Technology and Development* (Clarendon Press, 1975).

Sercovich, F. C., 'Foreign Technology and Control in the Argentinian Industry' (Ph.D. thesis, Sussex University, 1974).

Sharpston, M., 'International Subcontracting', *Oxford Economic Papers*, vol. 27 (March 1975).

Shetty, M. C., *Small-scale and Household Industries in a Developing Economy* (Asia Publishing House, 1963).

Singer, H., 'The Mechanics of Economic Development', *Indian Economic Review*, vol. 1 (August 1969).

Smith, A. D. (ed.), *Wage Policy Issues in Economic Development* (Macmillan, 1969).

Solow, R. M., *Growth Theory* (Clarendon Press, 1970).

Spencer, D. L., *Technology Gap in Perspective* (Spartan, 1970).

Staley, E. and R. Morse, *Modern Small Industry for Developing Countries* (McGraw Hill, 1965).

Stern, N. H., 'Experience with the Use of the Little-Mirrlees Method for an Appraisal of Small Holder Tea in Kenya', *Bulletin of Oxford University Institute of Economics and Statistics*, vol. 34 (February 1972).

Stewart, F., 'Transfer of Economic Concepts from Highly Developed to Less Developed Countries', in K. B. Madhava (ed.), *International Development, 1969*, (Society for International Development, Oceana, 1970).

Stewart, F., 'A Note on Social Cost-Benefit Analysis and Class Conflict in LDCs', *World Development*, vol. 3 (January 1975).

Stewart, F. (ed.), *Employment, Income Distribution and Development* (Frank Cass, 1975).

Stewart, F., and M. Stewart, 'Developing Countries, Trade and Liquidity: A New Approach', *The Banker*, vol. 122 (March 1972).

Stewart, F., and P. P. Streeten, 'Conflicts between Output and Employment Objectives in Developing Countries', *Oxford Economic Papers*, vol. 23 (July 1971).

Stewart, F., and P. P. Streeten, 'Little-Mirrlees Methods and Project Appraisal', *Bulletin of the Oxford University Institute of Economics and Statistics*, vol. 34 (February 1972).

Stewart, F., and J. Weeks, 'The Employment Effects of Wage Changes in Poor Countries', in Stewart (ed.) 1975.

Stigler, G. J., *Capital and Rates of Return in Manufacturing Industries* (National Bureau of Economic Research, 1963).

Stolper, W. F., 'A Note on the Multiplier, Flexible Exchanges and the Dollar Shortage', *Economia Internazionale*, vol. III (August 1950).

Stolper, W. F., and P. A. Samuelson, 'Protection and Real Wages', *Review of Economic Studies*, vol. IX (November 1941).

Strassman, W. P., 'Interrelated Industries and the Rate of Technological Change', *Review of Economic Studies*, vol. XXVII (October 1959).

Strassman, W. P., *Technological Change and Economic Development* (Cornell University Press, 1968).

Strassman, W. P., 'Mass Production of Dwellings in Colombia: A Case Study' (Geneva: I.L.O., mimeo, 1974).

Strathclyde University, *A Report on a Pilot Investigation of the Choice of Technology in Developing Countries*, (1975).

Streeten, P. P., *Economic Integration, Aspects and Problems* (Sythoff-Leyden, 1964).

Streeten, P. P., 'A Critique of Development Concepts', *European Journal of Sociology*, vol. II (1970).

Streeten, P. P., 'Technology Gaps between Rich and Poor Countries', *The Scottish Journal of Political Economy*, vol. XIX (November 1972).

Subrahmanian, K. K., *Imports of Capital and Technology. A Study of Foreign Collaborations in Indian Industry* (People's Publishing House, 1972).

Sunkel, O., 'National Development Policy and External Dependence in Latin America', *Journal of Development Studies*, vol. 6 (October 1969).

Sunkel, O., 'Transnational Capital and National Disintegration in Latin America', *Social and Economic Studies*, vol. 22 (March 1973).

Sylos-Labini, P., *Oligopoly and Technical Progress*, translated by E. Henderson (Harvard University Press, 1969).

Szentes, T., *The Political Economy of Underdevelopment* (Budapest Akademiai Kiado, 1971).

Timmer, C. P., 'Choice of Technique in Rice Milling in Java', and 'A Reply', *Research and Training Network* (September 1974).

Todaro, M. P., 'A Model of Labour Migration and Urban Unemployment in Less Developed Countries', *The American Economic Review*, vol. LIX (March 1969).

Todaro, M. P., 'A Theoretical Note on Labour as an "Inferior" Factor in Less Developed Countries', *Journal of Development Studies*, vol. 5 (July, 1969).

Turnham, D., *The Employment Problem in Less Developed Countries, A Review of the Evidence* (O.E.C.D., 1971).

United Nations Bureau of Economic Affairs, 'Problems of Size of Plant in Industry in Underdeveloped Countries', *Industrialisation and Productivity*, UNIDO, no. 2 (1959) and 'Plant Size and Economies of Scale', *Industrialisation and Productivity*, UNIDO, no. 8 (1964).

United Nations Conference on Trade and Development (UNCTAD), TD/B/AC 11/10, 'Major Issues Arising from the Transfer of Technology to Developing Countries', (1972).

United Nations Department of Economic and Social Affairs, *Panel on Foreign Investment in Latin America* (Colombia, 1971).

United Nations Industrial Development Organisation (UNIDO), 'Organisation and Operation of Cottage and Small Industries in Japan', *Industrialisation and Productivity Bulletin* no. 2 (1959).

UNIDO, ID/WG, 8/1, *Report of the Expert Group Meeting on the Selection of Textile Machinery in the Cotton Industry* (1968).

UNIDO, *Small Scale Industry in Latin America* (1969).

Vaitsos, C. V., *Intercountry Income Distribution and Transnational Enterprises*, (Oxford University Press, 1974).

Vernon, R., 'International Investment and International Trade in the Product Cycle', *Quarterly Journal of Economics*, vol. LXXX (May 1966).

Viner, J., *The Customs Union Issue* (Carnegie Endowment for International Peace, 1950).

Wagner, L. U., 'Problems in Estimating Research and Development and Stock', *Proceedings of the American Statistical Association*, Business and Economic Section (1968).

Weeks, J., 'Does Employment Matter?', *Manpower and Unemployment Research in Africa*, Centre for Developing Area Studies, Montreal vol. 4 (April, 1971); reprinted in Jolly *et al.* (1973).

Weeks, J., 'Wage Policy and the Colonial Legacy – a Comparative Study', *Journal of Modern African Studies*, vol. 9 (1971).

Weisskoff, R., R. Levy, L. Nisonoff and E. Wolff, 'A Multi-Sector Simulation Model of Employment, Growth and Income Distribution in Puerto Rico: A Re-evaluation of "Successful" Development Strategy', (Yale Growth Center, mimeo, 1973).

Wells, E. J., 'The Production and Marketing of Ballast in Kenya', mimeo, Institute of Development Studies, Nairobi (1970).

Wells, L. T., 'Economic Man and Engineering Man: a Choice of Technology in a Low Wage Country'. *Public Policy*, vol. 21 (summer 1973).

Wilber, C. K., *The Soviet Model and Underdeveloped Countries* (University of North Carolina Press, 1969).

Wilkinson, R. G., *Poverty and Progress* (Methuen, 1973).

Winston, G. C., 'Capital Utilisation in Economic Development', *Economic Journal*, vol. 81 (March 1971).

Winston, G. C., 'Capital Utilisation and Employment: a Neo-Classical Model of Optimal Shift Work', Research Memo no. 51 (Williams College, 1972).

World Development Movement, *Textiles – A Protection Racket* (1972).

Worley, J. S., 'The Changing Direction of Research and Development Employment among Firms', in *The Rate and Direction of Inventive Activity* (National Bureau of Economic Research, 1962).

Young, A., 'Increasing Returns and Economic Progress', *Economic Journal*, vol. XXXVIII (March 1928).

# Index